American Genius

American Genius

NINETEENTH-CENTURY

BANK LOCKS AND

TIME LOCKS

David Erroll & John Erroll | Photographs by Anne Day

The Quantuck Lane Press
New York

American Genius
David Erroll and John Erroll

Text copyright © 2006 by David Erroll and John Erroll.

All photographs copyright © 2006 by Anne Day except:

David Erroll: page 12, 17, 19, 21–22, 74 (right), 136, 152 (right), 170–71, 180–81, 186–89, 194, 197, 208, 217 (top left, top right), 226, 254–55, 267, 295, 300–04, 308–11, 314–17

John Erroll: pages 209 (right) and 270

Images courtesy of John Erroll: pages 14–16, 18, 26–28, 31, 34, 45, 54, 111, 122, 158–59, 164–65, 174, 266, 272, 274–75, 291

Schematics on pages 14 and 16 courtesy of Science Museum/Science & Society Picture Library (London, UK).

Images courtesy of General Society of Mechanics and Tradesmen: pages 9, 24–25, 118, 144 (left), 155 (right), 164, 242–43

Courtesy of Hall family: pages 29, 33

Courtesy of Dean Cross: page 209

Courtesy of James Shoop: pages 280–81, 288–89, 344

Images from United States Patent and Trademark Office: pages 44, 176, 198, 271. No copyright claimed on government works.

All rights reserved.
Printed in Italy.
First Edition.

The text of this book is composed in Adobe Caslon with the display in AT Medium.

Book design and composition by Laura Lindgren.
Manufacturing by Mondadori Printing, Verona.

Library of Congress Cataloging-in-Publication Data
Erroll, John.
   American genius : nineteenth-century bank locks and time locks / John Erroll and David Erroll ; Photographs by Anne Day. — 1st ed.
      p. cm.
   Includes bibliographical references and index.
   ISBN-13: 978-1-59372-016-2
   ISBN-10: 1-59372-016-5
  1. Locks and keys—United States—History—19th century.
  2. Locks and keys—United States—History—20th century.
   I. Erroll, David. II. Title.
TS519.5.E77 2006
683'.32—dc22   2006017746

The Quantuck Lane Press
Distributed by W. W. Norton & Company, Inc.
500 Fifth Avenue
New York, N.Y. 10110
www.quantucklanepress.com

W. W. Norton & Company Ltd.
Castle House
75/76 Wells Street
London W1T 3QT

1 2 3 4 5 6 7 8 9 0

This book is dedicated to the craftsmen and artisans
whose works of enduring beauty have remained all but unseen,
without whom this book would not have been possible;

and to

B.,

without whom this book would not have been probable.

CONTENTS

Acknowledgments 8

1. The American Bank Lock Industry
13

2. The Early Prominence of Key Locks: 1834–1856
37

3. The Combination Lock Comes of Age: 1857–1871
95

4. The Rise of the Time Lock: 1872–1888
143

5. The Era of Monumental Security: 1888–1899
239

6. Locks of the Early Twentieth Century: 1900–1915
297

7. Alarm Timers
327

8. Epilogue
343

Notes 345   *Glossary* 355   *Bibliography* 359   *Index* 362

ACKNOWLEDGMENTS

The art and science of technical endeavors survive only so long as people study them and pass them on. Safe and vault lock technology is no exception—in fact, it may be a science even more dependent upon its active practice for its longevity than most. Locksmithing was, until very recent times, an industry structured much along the lines of the guilds of the old world. Secrecy was paramount, not only for the commercial security of the locksmith's client but also for the financial security of the locksmith himself. The ability to help a locked-out banker or repair a malfunctioning vault door was highly valuable and not lightly taught to strangers.

With safe and vault technicians relying on confidence and secrecy to both maintain a professional reputation and protect themselves from competition, tradesmen generally shunned technical publication in favor of apprenticeship. An employee would be taken on and, after showing himself to be reliable and trustworthy, would be taught increasingly confidential details about types of locks and specific clients' equipment. Eventually the master apprentice would be made a partner in the business or strike out on his own, founding his own business, possibly taking some clients with him. As a consequence, the vast majority of technical skill and historical perspective was always just one generation from irrecoverable obscurity—a single generation of technicians uninterested in learning and teaching the details of the early locks was all that was required for the operation of these complicated mechanisms to be all but lost.

For a handful of the rarest locks it may be too late: too complex to be reverse engineered and too valuable to risk disassembly, their secrets seem to have been lost to time. For these, we can only marvel at the obvious details and wonder about the more subtle ones. Yet for the vast majority—even the rare and intricate—we can still gain an exceptional and refined understanding of the operations, the strengths, the innovations, and the weaknesses of these beautiful instruments, which, though replaced, have never been surpassed.

This is due in large part to a handful of people and we would be remiss if we did not acknowledge the unparalleled contributions to the understanding and preservation of safe and vault lock history by those dedicated to it. This volume would have been quite impossible without the following:

JOHN MALCOLM MOSSMAN was born on June 9, 1846, at 386 Greenwich Street in New York City, the eldest of the seven children of Malcolm Mossman and Christina A. Watson, who had both emigrated from Edinburgh, Scotland.

Always industrious and hardworking, Mossman left school at the age of twelve to work with his father, who had recently established a safe and lock business. Malcolm Mossman had been a foreman for the well-known safe maker Silas Herring, at West 13th Street and Ninth Avenue, but two years later, in 1860, Malcolm Mossman died, aged forty-five. John Mossman and his mother maintained the newly formed business for five trying years until agreeing to combine with another well-known safe maker, William Terwilliger. Shortly after, John was sent to the Chicago Works of Terwilliger & Co., and at the age of nineteen he was made foreman over two hundred workmen. When Terwilliger & Co. failed in 1876, young

Mossman returned to New York and found a business partner in Ira L. Cady, at 100 Maiden Lane, founding Cady & Mossman in 1877. Cady died in 1878 and Mossman continued alone, later bringing his brother William into the business. Mossman's business grew and soon included not only the buying and selling of safes, but also the sale, installation, cleaning, and care of the most complicated combination and time locks on bank safes and vaults throughout New York City.

Over the ensuing years, John Mossman parlayed his engaging personality, business acumen, and expertise in his field into a national following among bankers and bank architects. Gradually, he devoted an increasing portion of his time to the design and oversight of burglar- and fireproof bank vaults and safe deposit vaults. By the late 1880s his was considered the last word in such matters, and Mossman was overseeing projects for the largest banks and most important architects throughout the United States, including the Treasury Department, as well as numerous institutions in Europe.

In 1888, Mossman married Evelyn Angus, daughter of Thomas Angus and Catherine Farmer, who like Mossman's parents had emigrated from Scotland and had resided in New York since 1835. John and Evelyn had one son, John Malcolm, Jr.

Mossman's brother and business associate William died in 1902, and four years later John decided to retire. He divided his business into two parts, one to handle the sales of office and bank safes and the other to handle time lock maintenance contracts. Mossman turned over his safe sales business and its 55 Maiden Lane offices to the York Safe and Lock Company, a firm that would remain a significant factor in the safe industry for many decades. Mossman retained his time lock maintenance business, forming the J. M. Mossman Company of 23 Warren Street with Thomas F. Keating, a well-known manager from the Yale & Towne Mfg. Co. and a good friend.

On March 6, 1912, John Mossman died from a sudden onset of typhoid fever. Henry R. Towne, then president of Yale & Towne Mfg. Co., eulogized Mossman.

> I knew him for more than thirty years and had many opportunities to note and appreciate his technical skill in the field of work to which he had devoted himself. As a technical expert in vault construction and bank locks, he easily came first. His death closes the last chapter of a

John Mossman

record begun by Linus Yale, Jr. In his day, the latter was recognized as the leading expert in bank vaults and locks. Some twenty years before his death, Mr. Yale associated with his work in this field his brother-in-law, Ira L. Cady, who, after Mr. Yale's death, continued to do business as an expert in the field, especially among the banks of New York. Mr. Cady associated with him John M. Mossman, who later became a partner and who succeeded him in turn on his death. The record of business controlled by these three men is a long and honorable one. It has been my privilege to know them all, and I regret sincerely that the last of them is gone.

John Mossman's legacy included some of the largest and most notable bank vaults, many in New York but a number in other cities as well. The prevailing faith in Mossman's abilities was reflected in the clientele for whom he built vaults, a veritable who's who of nineteenth-century finance. As the supervising architect for the Clearing House Committee, he oversaw the construction of the New York Clearing House vaults. He installed a vault for the Mercantile Safe Deposit Company in the Equitable Life Assurance Society building that preserved several billions of dollars' worth of securities and property when the rest of the building was destroyed by fire in 1912. He also installed vaults for J. P. Morgan & Co., Mutual Life Insurance Company, Chemical National Bank, the New York Stock Exchange, Union Trust Company, Central Trust Company, Manhattan Trust Company, Fifth Avenue Bank, Bank of America, Bank of New York, Mechanics National Bank, Metropolitan Life Insurance Company, Washington Life Insurance Company, Bank of Montreal, and many others. Mossman also frequently consulted for the United States government as an expert, reviewing the safety and security of the vaults in the Treasury Department in New York and Washington, D.C.

Because he was in the business of bank locks and, particularly, time locks, Mossman was frequently called upon to replace older locks with newer ones. Over the years he increasingly recognized the ingenuity displayed in the previous century. As a result, he amassed one of the largest collections of obsolete, inventive, and custom-designed vault locks in the world. Further, many of Mossman's associates knew of his interest in these curiosities and his collection and were often happy to present him with unusual locks that had come into their possession.

Mossman had become a member of the General Society of Mechanics and Tradesmen of the City of New York in 1874 and had taken an active part in the organization, chairing the prestigious Entertainment Committee from 1882 to 1893 during the General Society's greatest popularity. After his retirement in 1902, he turned his focus to his lock collection, cleaning, fixing, and refinishing his most representative pieces. In 1903, Mossman presented his collection to the General Society, along with a generous endowment for its maintenance and exhibition, and founded the General Society's Museum Committee to oversee the General Society's growing collections. In keeping with John Mossman's wishes, the General Society makes the Mossman Lock Collection available for public viewing and research.

HARRY MILLER was one of the most dynamic, knowledgeable, and colorful figures in the history of American bank vault technology. His career began in 1924 at the age of twelve as a locksmithing apprentice at the Diebold Safe & Lock Co. in Canton, Ohio, where his father worked. Later, when his father was transferred to Washington, D.C., to install bank vaults, the Miller family moved as well, and soon Harry developed his expertise at opening the best then-current safes by manipulation. He was called upon to open dozens of safes for the United States military during the Korean War, a safe in the White House for President Franklin Roosevelt, and even a bullion chest for General Chiang Kai-shek, and soon he became known around the nation's capital as "Miller the Safe Man." Based on his understanding of the weaknesses of these safes, Miller secured fifty patents for improvements to safes and locks, including a manipulation-proof safe lock that led to his gradual acquisition during the 1950s of Sargent & Greenleaf, Inc., one of the most historically significant lock companies. As president of Sargent & Greenleaf, Miller moved the company from Rochester, New York, to its current home of Nicholasville, Kentucky, and after selling Sargent & Greenleaf he went on to found Lockmasters, one of today's most important educational institutions for the locksmithing trade.

Miller's locksmithing career bridged the trade's evolution from the time when substantially all of the industry's history was carried by word of mouth to its more modern model, and he documented many details of historically important locks that never would have been preserved otherwise. Harry generously taught and mentored all who sought his guidance and was single-handedly responsible for the preservation of much of the oral history of his field. Without his enthusiastic interest and guidance, we and other contributors would never have been in a position to assemble this work. Harry Miller died in 1998, leaving not only his extensive and important collection of bank locks, which is now on public display at Lockmasters in Nicholasville, Kentucky, but also the memory of a storied career of public service and individual achievment.

J. CLAYTON MILLER has made a number of significant contributions to the safe and vault industry. Clay rose through the management ranks of Sargent & Greenleaf, eventually named president in 1976, and, in 1981, he purchased Lockmasters from his father, Harry Miller. Clay recently changed the name of his company's Educational Division to Lockmasters Security Institute, offering a broad survey of technical courses ranging from locksmithing to security management. Clay has more than twenty patents to his name and continues to support the research and development section of Lockmasters. In the late 1980s, he founded the Safe and Vault Technicians Association, an organization that grew to include more than twelve hundred members around the world, eventually becoming part of the Associated Locksmiths of America.

Today, Clay Miller and his wife, April Truitt, operate a primate sanctuary on the grounds of their farm in Kentucky. Clay continues to maintain the extraordinary Harry Miller Lock Collection at Lockmasters in Nicholasville, Kentucky, and his generous assistance with access to both the lock collection and historical documents was instrumental to the foundation of many aspects of this book.

THOMAS F. HENNESSEY, SR., worked as a locksmith for many significant lock manufacturing companies in Connecticut and was the curator of the Lock Museum of America in Terryville, Connecticut, for thirty-two years. Besides holding a number of security patents, he is known for such achievements as developing the master keying system that was used throughout the World Trade Center in New York. Hennessey has acquired and preserved a wonderfully representative collection of locks as well as some of the most important historical catalogues and documents shedding new light on a number of areas of locksmithing history. Currently, he serves as the Lock Museum's curator emeritus, with his son having taken over the curator's office, and has given generously of his time, knowledge, and guidance to this work.

DANA J. BLACKWELL is a well-known and much respected horologist who has spent many years generously educating us in the fine nuances of clock and watch movements, details that have proven invaluable to a clear understanding of the time lock industry. He served as president of the Howard Watch and Clock Products Company in the 1970s and as director of the American Watch and Clock Museum of Bristol, Connecticut, into the 1990s. Thanks in part to Dana's insights into the Howard Watch and Clock Company's history and movements, today's view of the time lock industry is immeasurably clearer than it was only a few years ago. A respected author and lecturer on horological history, his internationally recognized level of expertise is reflected in his long friendship with Henry Stern, owner of Patek Phillipe. Today, Blackwell is retired and lives with his wife in Connecticut.

LYNN COLLINS is probably the most knowledgeable living authority on American safe and vault locks of the nineteenth century. A master machinist by trade, he has worked as a locksmith and vault technician and has taught and inspired us throughout our years of research and writing. With one of the most important bank vault key lock collections in the world, he has always been available and patiently helpful with his vast knowledge. Today, Collins is retired and lives in Florida with his wife, where he continues to collect and research English, American, and German key locking systems.

MARK BATES has worked in the lock industry for nearly twenty years, beginning as a locksmith and safe technician. In 1993, he founded the security firm MBA USA, Inc., which has grown to become a leading provider of security education and equipment.

Bates is a longtime member of the Associated Locksmiths of America (ALOA), the American Society for Industrial Security (ASIS), and the Safe and Vault Technicians Associations (SAVTA), and in 2000 he was inducted into the SAVTA Hall of Fame. He is the author of the reference book *Modern Safe Locks* and has contributed articles to leading industry journals.

In addition to his duties as an instructor at MBA, Bates has conducted classes for ALOA, SAVTA, and numerous regional associations. His teaching engagements have taken him throughout the United States and Canada, as well as to Austria, Denmark, England, Germany, Italy, Spain, and the Middle East. From these travels, he has developed a deep understanding of locks and security from a global vantage point.

Finally, we thank E. Rhett Butler, Dean Cross, Edwin and Elizabeth Fraser, Joseph L. Hall VI, T. Cartwright Hall, Orest Kalba, Joseph Kingsmill, Andrew and Daniel Kodas, James Shoop, and Steven Sommers for their thoughtful and generous contributions, which have helped make this book a reality. Any enumerated acknowledgment is, by its nature, incomplete. We would also like to thank those whose tireless assistance and unending dedication to the preservation of the history of the safe lock industry have made this work not simply possible but a success.

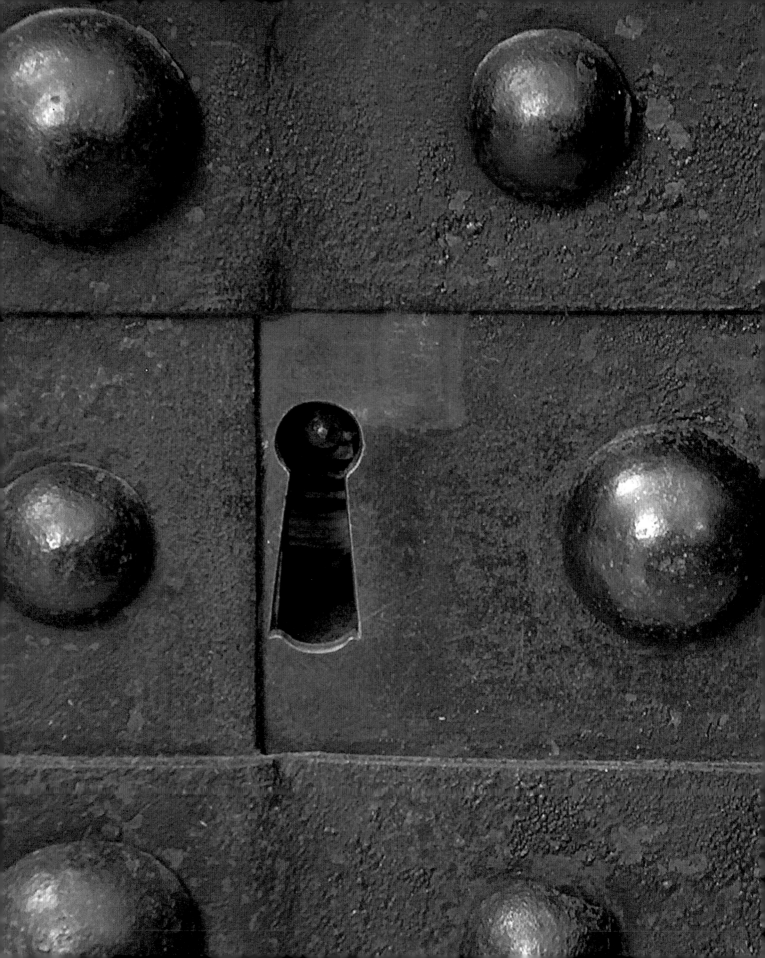

CHAPTER 1

The American Bank Lock Industry

Locks have long been valued by the knowledgeable and the novice alike, by some for their security and by others for their beauty and innovation. Masterpiece locks of fifteenth- and sixteenth-century Europe display marvelously intricate detail and engraving, serving as both the culmination of the apprentice's training and his first calling card as a locksmith of the guild. Almost unknown, on the other hand, are the masterpieces of American locksmithing, devoted to securing the safes and vaults of an expanding country. American bank locks of the nineteenth and early twentieth centuries are both examples of technical brilliance and jewels of artistry. They express the great pride their makers took in their craft and reflect the highest standards of ingenuity of their era. Yet all the while, such craftsmen knew that these works of art were destined to remain forever locked away from view.

Lock makers have long been a secretive group by intention. While proud of their work and willing to publicize their successes, they often were reluctant to have their most effective locks distributed broadly to the public. Even with the rigid guild system of medieval Europe long gone, the locksmithing profession retained much of this spirit, relying on apprenticeship and giving rise to a reclusive, fiercely competitive but tight-knit community. Consequently, for the public at large—and even for many collectors and museums—these locks have remained largely obscure.

Some of the historical detail of America's bank locks has perished forever, taken to the grave with a bygone generation of quietly studious technicians. However, John Malcolm Mossman's love of these mechanisms led him to preserve many of the treasures of this period, assembling one of the best collections of bank locks in the world. A careful overview of Mossman's collection and other important pieces combined with the first extensive analysis of period court documents can give us an understanding of some of the most peculiarly American technical innovations.

Prior to the nineteenth century, almost all locks of any importance in the United States were imported from England. The British empire had been shaped around the mercantilist theory that colonial holdings would provide the mother country with raw materials for domestic industry as well as markets for finished goods. The resulting British policy prevented skilled craftsmen of all kinds from leaving the country, including locksmiths. While this did serve to prolong American reliance on English lock making, even the best lock technology prior to 1775 was rudimentary. A banker or financier of the eighteenth century would have commonly stored bank notes and valuable documents in a wooden chest possibly reinforced with metal strapping or a heavy cabinet secured by an iron warded lock. Safes and vaults as we know them today—safes whose heft and toughness are defenses against theft—simply did not exist. It may be axiomatic to note that an undefeatable lock is of little comfort on an easily broken case, but the converse is also true: an impenetrable case is only as safe as the lock that secures it. With only nascent lock technology available, great expense was generally unjustified on the body of a safe.

The origins of warded locks are too obscure to be set forth with confidence, residing in the Dark Ages of medieval Europe,

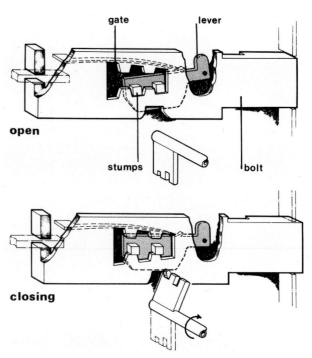

A schematic showing one of the two levers in Barron's 1778 lock. Both levers must be lifted correct distances for the lock to open. This lock was the harbinger of the first bank-quality locks.

but surviving examples are not uncommon. The warded lock employs a bolt thrown by the flange or bit of a key that is cut to avoid a sometimes intricate pattern of metal impediments or wards designed to prevent the key from turning. The warded lock is the quintessence of keeping the honest person honest—even in its heyday it suffered from a number of faults, not the least of which was that many were manufactured and widely sold that could be opened by the same key. Further, because a ward acts directly on the bit of the key, a blank key with an uncut bit covered in wax could be inserted and pressed against the wards, yielding an accurate impression of the actual key. The warded lock was also susceptible to a pick that could simply avoid the wards rather than negotiate them. And if these weaknesses were not enough, there is evidence that unscrupulous lock makers produced locks with keys cut to suggest an intricate pattern of wards but, upon inspection, were found to have few wards or even no internal defenses whatsoever. Keys designed with bits cut to avoid commonly placed wards, called "skeleton keys" because they had only the

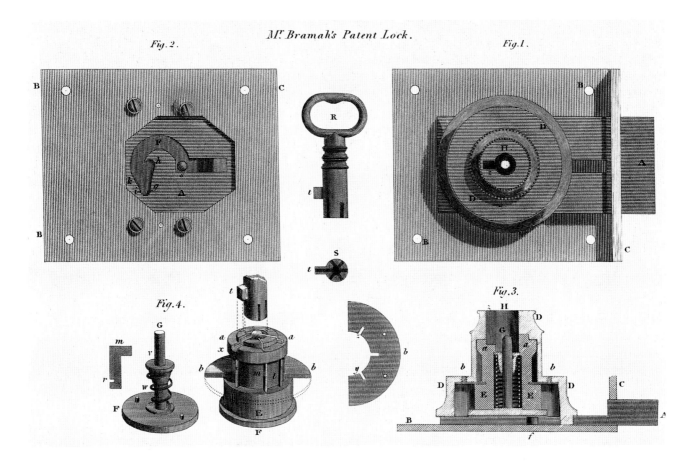

Joseph Bramah's 1784 lock. Bramah's key had a spur that allowed it to act as a turning wrench while six cuts of varying length in the end of the shaft acted as the bits. This design itself was unusual for its time, a factor that likely contributed to its success. Bramah's lock would remain one of the most secure locks for more than fifty years.

thinnest parts of the bit, were not uncommon, and at the height of the use of the warded lock, a savvy collection of twenty of these skeleton keys could open the majority of street doors in London.[1]

The dawn of modern lock design came with the English patent for an improvement on the lever mechanism, granted to Robert Barron in 1778. While a nascent lever mechanism was known at least as early as 1767, as described in the French publication *L'Art du serrurier*,[2] these lever-like locks had a single prototypical lever that required only that the key sufficiently lift it. This mechanism offered little more security than the basic warded lock. Barron's design retained a system of wards for security, but its main feature was a pair of dual-acting levers that required a key bit that would lift each lever to a specific height. With each lever held by a leaf spring, Barron's lock addressed all the inherent flaws of the warded lock: no simple impression of the lock's defenses could be taken; no single pick can actuate both levers simultaneously; and many unique key patterns could be made with inexpensive modifications to the levers.

In 1784 the first modern incarnation of the ancient technology of the pin tumbler lock was introduced in a patent awarded to Joseph Bramah. Pin tumbler technology itself was not new. In fact, pin tumbler locks were used by the ancient Egyptians at least as early as 1500 B.C.E. While the Bramah lock employs sliders rather than the pins of the Egyptian model, Bramah's sliders differ only in form. These notched sliders allow the cylinder to rotate when depressed, but they

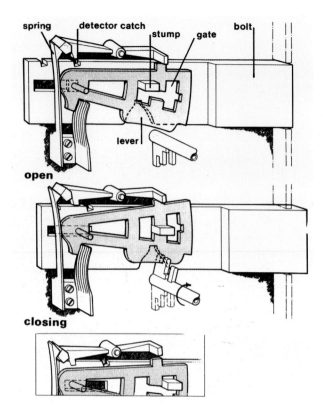

Jeremiah Chubb's lock design from 1818. Chubb improved upon Barron's lock, adding more levers, closer tolerances, and a false key detector catch. This detector catch (visible across the top of the mechanism) would lock the bolt if any lever was raised too high, alerting the owner that an incorrect key or pick had been used. Interestingly, Hobbs found that the detector catch made the job of picking easier for the true expert, as it offered a subtle increase in resistance that alerted him to the fact that the lever was too high.

also remain aligned and in contact with the tension spring without the aid of a surrounding block. Whether the rotation of the key is allowed by a notch or by a separation of a two-piece pin is ultimately a distinction without a difference and the success of Bramah's design was phenomenal. Bramah & Co. maintained a shop in Piccadilly, London, with a standing offer of 200 guineas (then £210, now worth over $20,000) for anyone who could pick a Bramah padlock—a challenge that would go unanswered for more than fifty years.[3]

The designs of Barron and Bramah were the standard for lock security and remained substantially unimproved until the 1818 lever design patented by Jeremiah Chubb. Chubb used a six-lever design similar to Barron's two-lever lock, but he added a detector catch that would block and hold the bolt if any lever was raised too high, whether by picking or by the use of a wrong or false key. A subtler but not insignificant aspect to the Chubb lock is that if the detector catch is thrown, the next time the true key is used the correct key must relock the bolt before the lock can be opened, alerting the user to the failed attempt.

With this new measure of security, the Chubb lock joined the Bramah lock in the public's esteem as simply unpickable. The rewards for picking the Bramah and Chubb locks stood unclaimed until 1851, when American locksmith Alfred C. Hobbs arrived in London. With an innovative set of picks, Hobbs opened Chubb's lock in twenty minutes.[4] While there was some controversy over Hobbs's defeat of the Bramah lock, the company did finally pay Hobbs the 200 guinea prize under protest, insofar as he had indeed opened the lock. So remarkable were Hobbs's feats that the *Times* of London noted:

> We believed before the exhibition opened we had the best locks in the world, and among us Bramah and Chubb were reckoned quite as impregnable as Gibraltar—more so, indeed, for the key to the Mediterranean was taken by us, but none among us could penetrate into the locks and shoot the bolts of these makers.[5]

This was the beginning of the growth of safe and vault technology, which would be a coevolution of two distinct sciences: the design of the body of the safe to prevent violent destruction and the complication of the locking mechanism to prevent picking or cracking. With the development of what were then considered high-security key locks, the safe-making industry began a transition away from wood and wrought iron toward forged iron and steel, with the appearance of strongboxes that posed significant hurdles to the casual criminal intent on forcing the container open. This introduction of the steel strongbox was the first salvo in what would become a running battle between the safe maker and the safecracker. This thief vs. banker competition would continue for more than a century, until major changes in the United States banking and finance system instituted during the Civil War ultimately spelled the end of banks' use of gold and privately issued banknotes.[6]

The steel strongbox served as a safe in the early part of the 1800s and in remote areas. Weighing from eighty to two hundred pounds, a strongbox could be quite difficult to move when loaded, but they were universally made with handles. Since they were designed to be carried, their security was inherently limited. This example was made around 1850. (Courtesy of E. Rhett Butler.)

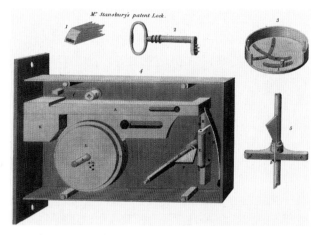

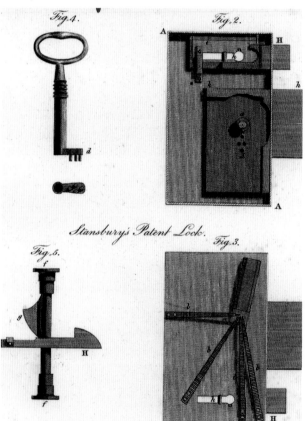

Stansbury's lock of 1807. The earliest known American design, Stansbury's lock relies on a pattern of pushpins to release its cylinder. Unlike proper pin tumbler mechanisms, Stansbury's lock doesn't require the pins to be a specific length, making it unfit for true burglarproof safes. This basic design was often supplemented by a pattern of wards.

While the strongbox commonly featured a relatively secure lock and sufficient mass to be unwieldy (a fully loaded strongbox could weigh in at more than two hundred pounds), strongboxes were in fact designed to be transported, featuring handles and only modest weight of their own. This presented a strategic weakness: the prepared criminal could abscond with the strongbox and force it open at his leisure. It became clear that immobility was a major line of defense for the safe maker, requiring the criminal to force or crack the vault in place, risking discovery. Over the next fifty years, safes and vaults would evolve into the inertial behemoths we know today.

While early American lock makers copied well-known English designs, advances in safe locks soon began in the United States. In 1807, A. O. Stansbury patented an adaptation of a door lock closely based on the Egyptian pin tumbler design. Stansbury's mechanism offered good security for its price, and locks based on Stansbury's patent were made and used even as late as the 1840s in commercial document safes such as the hobnail safe of Jesse Delano. Delano produced this hobnail safe from 1826 and amassed quite a fortune through its popularity over the ensuing twenty years. Though most commonly used by individuals and small business proprietors, the formidable-looking Delano hobnail safe was used for a time by the First State Bank of Shawneetown, Illinois Territory.[7] A rare safe today, the Delano hobnail safe in the Mossman collection is one of fewer than ten known surviving examples. Made around 1840, this example is in particularly fine condition. Despite its popularity, the Stansbury lock could not approach the complication, security, and renown of its English counterparts.

The first American-made lock sufficiently secure to be a true bank lock is thought to have been patented by J. Perkins on March 23, 1813 (information probably obtained from the donors).[8] However, the actual patent has not been found, likely due to the 1836 fire at the United States Patent and Trademark Office. It does not appear among patent papers recovered from nonexhaustive British sources. The Perkins bank lock combined elements of the pin tumbler lock and what would later come to be known as a combination lock. The internal bolt-stopping arrangement is essentially a pin tumbler with three three-piece blocks held across the lock case and bolt by leaf springs. The modern pin tumbler mechanism developed by Linus Yale, Jr., is foreshadowed by how the Perkins lock

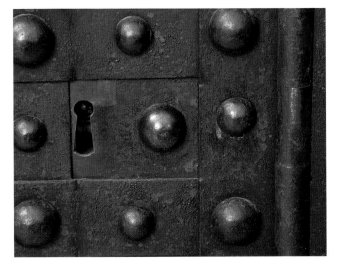

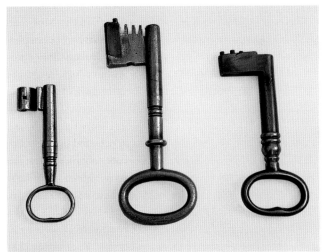

Hobnail safes were popular for their affordability and imposing appearance. This Jesse Delano model dates to about 1840 and is mostly wood covered with metal strapping. It features a warded version of the Stansbury lock and a keyhole that can be hidden by a false hobnail. This safe was best suited to keeping honest people honest, as its construction and lock would have been little match for a skilled thief.

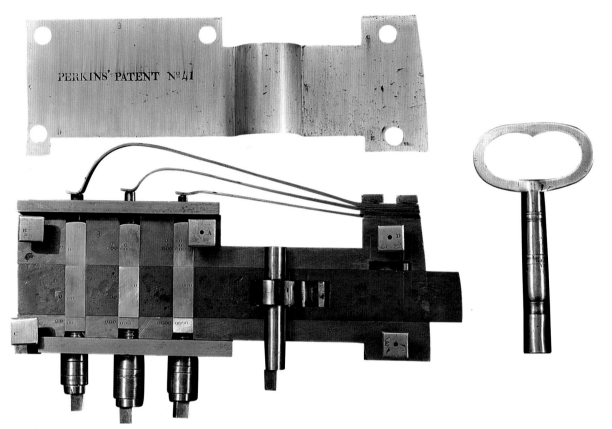

Perkins's bank lock was patented in 1813 and combined nascent key and combination lock principles, offering far better security than its contemporaries. However, at this early point, safes and vaults had yet to be made that could match the Perkins lock's protection. Ahead of its time, difficult to use, and likely expensive, it was not widely made or employed.

aligns strategically placed breaks in the pins with the abutting line of a moving bolt. The method of operating the Perkins lock, however, is the precursor to the modern combination lock. Early in the development of locks operated by manipulation rather than by key, makers used the term "permutation lock" alongside "combination lock." The term "combination lock" has taken firm hold in the lock lexicon, although the astute mathematician will quickly note that a true combination lock could be opened by dialing the correct digits in any order, while in fact all such locks require the digits be dialed in a specific order, properly termed a permutation. The last safe lock patented as a "permutation lock" was Milton Dalton's 1878 lock used in the Dalton Consolidated Triple Guard.

While the first lock easily recognizable as a combination lock is appropriately credited to Butterworth, Perkins's lock relies not on the simple possession of a key but rather on the execution of a preset series of turns on the controlling bolts.

This union of key- and combination-lock principles created a bank lock offering security noteworthy for its time. Even so, vaults as we know them today were not yet common in the banking industry and Perkins's lock, ahead of its time, did not find widespread adoption. The Perkins lock was installed and used for many years in a vault at the National Mechanics and Traders Bank of Portsmouth, New Hampshire, which made a gift of it to J. M. Mossman after it was removed.[9]

The Second Bank of the United States was chartered in 1816 and a building to house it was constructed in Philadelphia between 1819 and 1824. This building on Chestnut Street later served as a customhouse from 1845 to 1935 and is now part of Independence National Park. The original lock from that vault was made by Joseph Nock, father of the later well-

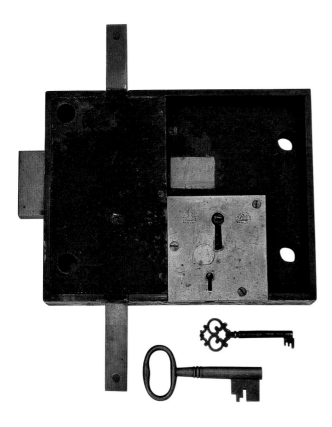

Possibly made as early as 1824, this vault lock by George Nock is the first American lock to use a smaller lock to guard the main keyhole. Its patent may have been for its lever mechanism or possibly its anti-gunpowder feature—a simple hole in the bottom of the main lock case.

The end of the bolt is embossed with Nock's name and a federal-style eagle common in the early United States. Though it claims the design was patented, the papers were likely lost during a fire in 1836.

known padlock maker George Nock. The bolt-end embossing claims a patented design but no corresponding patent papers remain, suggesting that the patent was filed before 1836 and lost in the Patent Office fire that year. Its embossed medallion also shows a federal eagle, common in the 1820s and 1830s, indicating that this lock could have been made for the initial vault installation in the early 1820s. This lock is the earliest known American vault lock to feature a secondary lock to control a primary keyhole: one key throws back a slug blocking the main keyhole and a second key throws the main bolt. The smaller secondary lock is brass throughout except for the blocking slug, which has a fitted steel insert. Its five-lever mechanism uses a barbed fork-gate design thought to be unique, and possibly the subject of the unidentified patent. The patent may also have been granted for the two antipowder devices, the blocking slug for the primary lock, and the hole in the case bottom in the secondary lock. The primary lock's four-lever mechanism is shown on page 22. All aspects of this lock are finely made with brass leaf springs and a steel case.

At the 1928 publication of *The Lure of the Lock*, nothing was known of the lock shown on page 23, the details surrounding it having been lost with the death of John Mossman. Finding it quite a mystery, Hopkins described it only as:

> *Odd Key Lock.* This is an extraordinary lock, and seems to depend for its security on the tension of the springs which seem to be controlled by pins turned by wrenches, while the bolt proper is shot and retracted by a wrench rack and

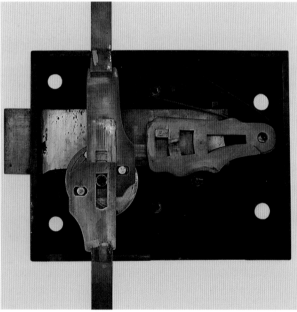

TOP: *Nock's secondary lock uses a barbed fork-shaped lever gate found nowhere else.*

BOTTOM: *The interior of Nock's lock shows the sleeves and the rotating disk geared to the vertical bolt extensions. All three bolts are retracted simultaneously when the lock is opened.*

pinion. We have used due diligence in trying to find out how the lock operates, but without success.[10]

However, a diligent search of patent records lost during the Patent and Trademark Office fire of 1836 but subsequently recovered from concurrently kept British documents[11] in 1999 revealed that this lock was patented by G. A. Rodgers in 1830.[12] Similar in principle to the Perkins lock, Rodgers's lock adds a fourth locking block and replaces leaf springs with coil springs.

Many safe and vault locks were patented in early-nineteenth-century America,[13] but only a small subset of the earliest models were installed in even one bank vault. This is the only known example of the lock, and, although it may have been installed in a vault, efforts to clean and refinish the lock in the early 1900s have unfortunately destroyed any evidence. Without new information coming to light, Rodgers's patent cannot be elevated from the great mass of good ideas to the cadre of the exceptional and will continue to rely on singularity for its interest.

No vault is wholly impervious to theft. A safe or vault can, however, be made impossible to crack before the owner's regularly scheduled return and without arousing alarm. By the 1830s safes had started to take on the formidable heft required to prevent thieves from absconding with the entire safe and handles, such as those on the Jesse Delano hobnail safe, which became rare outside of strongboxes intended for transport. Consequently, criminals' focus shifted to either the manipulation of the lock or the destruction of the safe. Until this point, the threat to bank locks from explosives was relatively limited. The only substance available was black powder, a weak explosive relative to the strength of metals available to lock makers. Consequently, solid construction and a few strategically placed holes in the bottom of the case could render a bank lock impervious to powder. Then Ascanio Sobrero's 1846 discovery of nitroglycerin and Alfred Nobel's subsequent stabilizing of it into dynamite brought forth the first true high explosives, substances that detonate with a supersonic shock wave that generates a self-sustained cascade of pressure-induced combustion. These posed a far greater threat to safes—especially nitroglycerin, a liquid that could be introduced or "loaded" into the gap between the door and the jamb, attacking the safe at its weakest point. The archives of the Lockmasters

Rodgers's 1830 vault lock. Based on a principle similar to Perkins's design, Rodgers's lock included a geared mechanism that allowed the tumblers to sit upright while being turned by the four square arbors visible to the left of each tumbler, reducing the case depth.

Security Institute contain a description by a vault technician from this era describing how nitroglycerin is made and used to blow open a safe, with specific attention to detecting an unexploded charge left by an interrupted safecracker and the danger associated with defusing such nitroglycerin loading. The highest-quality safes and locks from this point on would need both complications and strength.

Among the first generation of American lock makers was Solomon Andrews, a renaissance man who possessed a diversity of inventive talent possibly second only to Thomas Edison. Andrews designed, built, and flew the first dirigible aircraft in the United States in 1863 and, unlike the balloonists before him, he seems to have had success in not only surviving his flight but also reasonably piloting his course despite his design being motorless, relying on balloon, sail, and rudder. In all, Andrews is credited with twenty-four inventions, including improvements in barrel-making machinery, sewing machines, fumigators, forging presses, bicycles, gas lamps, coal kitchen ranges, and a wickless oil burner.

Despite his inventive nature, Andrews's public life is most commonly understood as one of public service. A physician by trade, Andrews had a medical practice in his hometown of Perth Amboy, New Jersey. He served as the president of Perth Amboy's Board of Health and was elected to three terms as mayor. He later returned to the Board of Health during two trying periods when the town was in the throes of cholera and yellow fever outbreaks and oversaw the construction of Perth Amboy's first sewer line. Andrews served as collector for the Perth Amboy port, councilman, justice of the peace, and, combining his penchants for invention and public service, he founded an institute for inventors. But Andrews does not seem to have been an entrepreneur, and while there is no indication that he suffered any particular financial setbacks, he never turned his innovative lock designs into the lucrative business that it might have been. Rather, his locks seem to have been made one at a time, likely on request. In fact, no

Linus Yale, Sr. (1797–1857) [left], and Linus Yale, Jr. (1821–1868) [right], were two of the most creative and influential designers of American bank locks.

two known surviving Solomon Andrews locks are exactly the same, with differences ranging from the subtle (case size) to the salient (number of levers). This may seem shortsighted today, but one should bear in mind that, even by 1840, there was little indication that bank security could ever support the large-scale lock production that would develop over the next fifty years.

Unlike Andrews, Linus Yale, Jr., made lock making his primary business and his insight and creativity would form the basis for much of what we consider basic principles of locks today. Born in Salisbury, New York, in 1821, Linus Yale, Jr., first set out to be an artist but eventually began working in the Newport, New York, lock shop of his father, himself a successful bank lock maker. Yale received his first lock patent in 1851, and over the next twenty years he would be awarded fifteen lock patents and go on to invent his brilliant Magic Bank Lock and Double Treasury Lock.[14]

After his father's death in 1857, Yale moved to Philadelphia and finally to Shelburne Falls, Massachusetts, focusing on designing and producing some of the highest-security locks of the day. Yale's first firm in Shelburne Falls was a partnership with financial backer Colonel Halbert Greenleaf, and while Yale & Greenleaf operated for only a short period, it established both Greenleaf in his important role as financier in the lock business and Yale in his role as designing genius. After the dissolution of the firm, Yale moved his business to Stamford, Connecticut, founding Yale Lock Manufacturing Company with engineer and financial backer Henry Towne in 1868. Linus Yale, Jr.'s death on Christmas Day later that year left Towne with the reigns of the company, and under his direction it would become and remain a major force in the bank lock market.

After the partnership between Greenleaf and Yale ended, Greenleaf moved to Rochester, New York, and founded Sargent & Greenleaf with James Sargent. Born in 1825, James Sargent was a Rochester native and he lived his first eighteen years on his parents' farm. After getting a job in a woolens factory, he became known for his expertise at machinery maintenance and oversaw an extensive weaving room until 1848. Sargent spent the next four years as a daguerreotypist, traveling the country and succeeding financially. He returned to the Northeast and formed the firm Sargent & Foster in Shelburne Falls, Massachusetts, producing a patented apple corer, and for five years was a significant success. Unfortunately, a financial crash in 1857 combined with the near total failure of the 1857 and 1858 apple crops spelled the end for Sargent's fledgling company.[15]

Familiar with fellow Shelburne Falls businessmen Yale and Greenleaf, Sargent became a traveling representative for their locks, in the process learning how to pick his competitors' locks. Soon Sargent could open Yale's locks as well and set out to design his own bank lock. James Sargent became a significant figure in the bank lock market with the simultaneous introduction in 1865 of his Micrometer, which could help open the best combination locks of the day, and his new Magnetic Bank Lock, which was impervious to the Micrometer. It was the enormous popularity of this Magnetic Bank Lock and its improved successor, the Automatic Bank Lock, that led to Sargent's 1867 partnership with Greenleaf and return to Rochester, where Sargent & Greenleaf would remain until 1974.[16]

The advances in safe construction and combination lock design during the 1850s and their widespread adoption during the 1860s forced bank robbers to resort to more drastic means. Some turned to the unreliable and dangerous explosives and cutting torches of the day, but others adopted a tactic that would become known at the time as a "masked robbery."[17] Far different from the image we have today of bandana-masked desperadoes bursting into a frontier bank, the first generation of masked robberies were nighttime affairs and the largest were generally the subject of news pamphlets detailing the crimes, considered fantastic in their day. One of the best recorded was known alternatively as "The Great Burglary" or the "Astounding Narrative of a Masterpiece of Criminal Ingenuity," the 1876 robbery of the Northampton National Bank, Northampton, Massachusetts.[18]

The so-called Great Burglary began not at the bank but soon after midnight at the home of John Whittelsey, a bank cashier, two-thirds of a mile away. Seven robbers, one for each

James Sargent was a brilliant bank lock inventor, introducing designs that set standards for security in the mid-1800s.

resident, burst into the house and tied up the Whittelsey family. The robbers demanded that Whittelsey divulge the combination to the bank's safe, and when he gave them the numbers they diligently took them down. After a short period, they demanded that Whittelsey repeat the combination, but having given a false one made up on the spot, he could not recall it now. The robbers choked Whittelsey and, according to one account, "punched him in the chest with a large lead pencil."[19] The true combination obtained, the robbers made off with more than $500,000 in banknotes, stock certificates, and bonds. At least three of these perpetrators were later captured and all the money recovered,[20] but the message to the banking industry was clear: with the newest technology in safe construction and combination locks, the weak point in the security system was now the human element. Bankers would need a lock that could keep out not just the robbers, but the bankers as well.

James Sargent was among the first to design a commercially successful time lock. He is seen here on a broadside advertising a prototype of his Model 2 time lock. This early Sargent & Greenleaf advertisement may precede even the firm's earliest time lock sales, since such publications commonly listed client banks.

Although the idea of a time lock had been around since at least 1831, with a patent awarded in England to Rutherford,[21] banks had been sharply reluctant to adopt them. Even when working properly, the time lock barred both the welcome and unwelcome alike—if it failed, the safe or lock would have to be at least partially destroyed. The earliest time lock to be commercially used is now known to have been Holbrook's Automatic Lock of 1858, but it was the masked robberies of the 1870s that catapulted the time lock to the forefront of bank security. By then Holbrook and their Automatic were long gone, but a new generation of time locks was about to come into prominence.

The mid-1870s saw a flurry of time lock invention with models from many makers offered for sale, many based on the small, accurate, and reliable clock movements of the E. Howard Clock Company. The largest of these E. Howard–supplied makers was the Yale Lock Mfg. Co. with its Double Pin Dial time lock. Yale Lock's main competitor was Sargent & Greenleaf, which made all its own clock movements at the Rochester factory. Early on, patent priority among time lock makers was unclear, with many different individuals claiming broad patent rights based on any small improvement. Both Sargent & Greenleaf and Yale seemed to have patents by purchase or assignment that could have given either company absolute priority. But when both Sargent's and Yale's reissued patents were held valid during a patent interference action,[22] Sargent and Yale realized that they would likely have to resort to the courts to ascertain who owed whom royalties and in what amount.

Then in 1877, facing lengthy and expensive litigation, Yale and Sargent reached a watershed agreement that would completely change the landscape of the time lock industry. Possibly realizing that neither party was likely to prevail totally in court and have the time lock market to itself, Yale and Sargent entered into a contract that, by today's standards, would qualify as a "smoking gun" violation of federal antitrust law.[23] However, in their day this was nothing more than savvy business practice. Under this new "Contract Respecting Time Locks,"[24] Yale and Sargent would not only settle their legal claim, but also work together to divide and dominate the time lock industry.

This contract pooled Yale's and Sargent's time lock interests and provided for a number of other changes in their relationship. First, Yale and Sargent agreed to submit their legal claims to binding arbitration, setting forth the rules and agreeing to limit their liability to $50 per time lock or $50,000 in total[25]—understandable goals, legal even today. However, the agreement went on to fix wholesale and retail prices[26] and divide the country between the two companies,[27] with one company's agents acting for both makers impartially. Other sections pooled the two companies' patents, fixed the royalties demanded from users of "infringing" time locks,[28] and agreed to share equally the costs and responsibility for litigation of these pooled patent rights. Further, any future competition between Yale and Sargent was completely stultified by agreements to fix the safe makers' discount at 20 percent, to catalogue their time lock sales and split the resulting profits equally,[29] and not to introduce new models for five years without the other party's prior consent.

In the end, this agreement seems to have benefited Sargent far more than Yale. Only Sargent introduced a new time lock during the five-year moratorium, but whether Yale sought permission and was refused we cannot know. The arbitration outcome favored Sargent as well, with the arbitrators finding all the patents and reissues in question valid. Consequently, Yale was found to owe Sargent amounts ranging from $10 to $40 for each Yale Pin Dial sold both before and after their agreement due to Sargent's reissue no. 7,947, offset by $20 per time lock sold by Sargent based on Yale's ownership of Samuel A. Little's reissue no. 7,104 bought from Ely Lillie.[30] Since the arbitrators premised all their awards on reissued patents, the outcome may have been much different had the parties continued in court, as suggested by the ultimate outcome of *Yale Lock Mfg. Co. v. Berkshire National Savings Bank*. But that decision by the United States Supreme Court would not be handed down until May 5, 1890,[31] over twelve years later.

While the Yale and Sargent time lock interest pooling agreement changed the way these two major companies did business with each other, a greater effect was felt by the rest of the time lock industry. As stipulated by their agreement, Yale and Sargent began to aggressively pursue all other time lock makers, using the combined might of their patent portfolios to snuff out other competitors. One interesting facet of these patent suits was that, although their ultimate target was other makers, Yale and Sargent only ever named banks as defendants, a tactic that would prove quite effective. The banks were sued as "users of infringing locks," and while a judgment could effect only the one bank, the industry was small and word spread quickly. Banks in the market for a time lock would be prudently wary of laying out hundreds of dollars for a model that could land them in court, force them to pay a licensing fee of $200 to $250, or surrender the lock completely. Further, banks were relatively soft litigation targets even after they had a competitor's time lock in place. The banks were not interested in the validity of Sargent's and Yale's underlying patents, but rather in security—security equally well provided by a Yale or a Sargent time lock as by a competitor's. Consequently, the banks were usually keen to reach a settlement, even for $250, especially after Yale and Sargent had a series of court victories to show.

These lawsuits began in 1879, and the first to be decided were a pair of suits against the Norwich National Savings Bank and the New Haven Savings Bank for use of time locks made by Holmes, tried before Judge Nathaniel Shipman. Holmes seems to have overseen the defense of these banks since his company, Holmes Burglar Alarm & Telegraph, had committed to refund the purchase price should it be enjoined from using them by the court. Such agreements were common, since prior to 1879 many companies had advertised patent priority. However, the new suits filed by Yale and Sargent jointly were the first to force these smaller companies to make good on their pledges, pledges that turned out to be quite expensive. Unfortunately, Holmes had little success in court and decisions in these two cases were handed down in March

Gentlemen: In consideration of your purchase of a SARGENT & GREENLEAF TIME LOCK, and upon condition that the lock receives fair treatment, is, at your expense, examined, and cleaned and repaired if necessary, once each year, by an expert to be designated or approved by us (the charge for whose services, however, shall not exceed $10 for each such examination and cleaning,) and is used according to the printed directions annexed hereto and constituting a part of this agreement, we guarantee, for and during three years from this date, the following, to wit:

First, The proper working of the said lock, and

Second, That, in the event of its locking you out by reason of any fault or defect in the lock, we will pay all expenses of opening and repairing or restoring the door; provided, however, that we are promptly notified of any previous irregularity in the working of the lock, and that the use of the lock be suspended in the event of any such irregularity occurring, until it has been remedied; and provided, further, that, should you ever be locked out by said lock, we shall be at once notified thereof by telegraph, and shall be allowed not less than _____ hours from the receipt of such communication in which to commence (by a person or firm to be designated by us) the opening of the door on which said lock is in use; but it is agreed that the work of opening said door shall be begun with all possible dispatch after our receipt of the notification aforesaid.

Rochester, N.Y. _____ 187__

Not valid unless countersigned by _____

Countersigned at _____ 187__

_____ Agent.

In consideration of the purchase of a SARGENT & GREENLEAF TIME LOCK,

by _____

we, Sargent & Greenleaf, of Rochester, N.Y., and The Yale Lock Manufacturing Company, of Stamford, Conn., do hereby agree to warrant, defend and hold harmless the said _____ from all damages and expenses on account of any suit or suits for the infringement of any Patent or Patents, now existing or that may hereafter be issued, which may cover or be alleged to cover any feature in the construction of said lock, or in our method of applying said lock in connection with the bolt-work or other locks of the door on which it is placed; and if any such suit or suits should be brought, we agree to assume the defense thereof by our own counsel, and at our own cost; provided, however, that we are both promptly notified of the service of legal notice of any such suit upon said _____

THE YALE LOCK MANUFACTURING CO.

By _____

New York, _____ 187__

Not valid unless countersigned by _____

Countersigned at _____ 187__

_____ Agent.

With their time lock interests pooled, Yale Lock Mfg. Co. and Sargent & Greenleaf issued sales agents unified sales and service contracts. Their noncompetition agreement would probably run afoul of federal antitrust law today, but in 1877 this was simply considered good business.

of 1881,[32] with Shipman finding the Yale and Sargent's reissues to be valid and infringed by Holmes's time lock design. These decisions would be referred to in many subsequent decisions in other federal circuit courts and soon gave rise to a solid body of law supporting Yale's and Sargent's dominant position in the time lock market.

Along with their assault on smaller time lock makers in the Northeast, Yale and Sargent filed suit in January 1879 against a bank based on its use of a time lock made by Joseph Hall, an individual who would prove to be a much tougher adversary. Hall had been an early and successful entrant into the bank security business but has remained a surprisingly obscure figure. Born in Salem, New Jersey, on May 9, 1823, Hall moved with his family to Pittsburgh, Pennsylvania, in 1832. Working from the age of eight, Hall was employed in the river steamboat business from age seventeen to twenty-three. In 1846 Joseph Hall and his father formed a partnership in the then-emerging industry of safe manufacture, and in 1848 they moved this business to the more important commercial center of Cincinnati, Ohio.[33] This partnership, like most, would go through many incarnations as minority partners and interests changed: specifically, E. & J. L. Hall (1846–1850); E. Hall & Co. (1850–1851); Hall, Dodds & Co. (1851–1852); Hall & Dodds (1852–1854); Hall, Dodds & Co. (1854–1856); Hall & Carroll (1856–1858); Hall, Carroll & Co. (1858–1862); Joseph L. Hall & Co. (1862–1867). Joseph Hall's business acumen was clearly illustrated during the Civil War, when demand for safes all but disappeared. Without machinery or experience, he undertook a government contract to alter and rifle five thousand Austrian muskets in thirty days.[34] Hall made good on this and other government contracts, and by 1867 he was the majority stockholder, president, and treasurer of Hall's Safe & Lock Co.,[35] the name his company would keep for more than thirty-five years. With the end of the Civil War and the transition of the economy from a war footing back to commerce, the market for safes and bank locks recovered, and though Hall's Safe & Lock held a significant market position, it had yet to set itself apart from the competition. That would change in the span of two years.

In 1869 Hall employee Henry Gross was granted a patent for a design that would serve as the company's flagship combination lock for years to come, the Hall Premier. The Premier made Hall's Safe & Lock a formidable competitor for

Joseph Hall (1823–1889) was a major force in the safe and lock industry for nearly forty-five years. (Image courtesy of the Hall family.)

other important lock makers, such as Yale & Towne Mfg. Co. and Sargent & Greenleaf, yet Hall's company would not gain broad public renown until 1871. Headquartered in Cincinnati, Hall's Safe & Lock was a convenient choice for safes for much of midwestern and frontier commerce, including the burgeoning metropolitan center of Chicago. Then, for two days in October, the Great Chicago Fire of 1871 tore through the city.[36] What was a disaster for the city was a boon for Hall, who had sold about five hundred Hall safes in Chicago. Hundreds of testimonials surfaced about Hall safes that had withstood the fire, protecting their contents sometimes for more than a week in burning rubble.[37] This news brought in a flood of new business and by 1872 Hall had installed the largest fireproof vaults in the country in Pittsburgh and New Orleans and was reported to have a factory four times the size of his competitors', employing five hundred mechanics and producing twenty-five safes per day.[38]

By the early 1870s, as safes and combination locks became increasingly refined, patent protection and successful patent litigation were taking on a greater importance in the safe and lock industry. Although Joseph Hall had not yet been party to such litigation, he was certainly aware of Sargent & Greenleaf's successful lawsuits against George Damon's Bank Lock Company and Yale. With this in mind, Hall employed Milton Dalton in 1873 to pursue what must be one of the boldest and broadest acts of corporate espionage ever. Dalton set out to interview every active safe maker in North America and Europe posing as a writer researching a book on safe construction. In the process, he managed to obtain signed and sworn statements from employees of many major safe companies detailing the exact nature and development of their safe making techniques.

The resulting book, prodigiously titled *History of Fire & Burglar Proof Safes, Bank Locks and Vaults in America and Europe—Useful Information for Bankers, Business Men and Safe Manufacturers*, was published in a run of only sixty-six copies and never distributed,[39] but its contents did establish that Hall held the earliest patented use of conical bolts in safe manufacture. This evidence helped Hall prevail in a patent infringement suit against well-known safe maker MacNeale & Urban Co. of Hamilton, Ohio.[40] MacNeale & Urban survived this loss and continued in business for nearly thirty more years, until 1903, but was never the major threat to Hall that it might have become. (Two copies of Dalton's *History* survive today, offering one of the most reliable references on nineteenth-century safe making.) From this point on, Milton Dalton would play an increasingly important role at Hall's Safe & Lock, eventually becoming general manager. However, it soon became clear that Dalton's investigative and administrative contributions would be roundly eclipsed by his extraordinary inventive genius. Between 1873 and 1890, Milton Dalton was awarded many patents for combination and time locks, ranging from the elegantly simple to the surprisingly complex, and he would remain a major inventive force at Hall's companies until the 1890s.

Joseph Hall recognized the necessity of both inventive creativity and legal protection, and in a farsighted move he spun off his time lock business into its own company. While a number of early and experimental models of time locks were made and labeled "Joseph Hall," Hall realized that the budding time lock industry was a perilous one, from both regular market forces and litigation. Patent litigation had already spelled doom for some lock makers, including Lewis Lillie and even the once robust New Britain Lock Co., maker of the Pillard time lock. Hall insulated his established and lucrative safe and combination lock business from these threats, founding the Consolidated Time Lock Company in January 1880.

Though Hall was willing to take precautions, he was no shrinking violet. Hall's motto was once reported to have been "Let me alone and I will let you alone; but tread upon me and I will strike,"[41] and when Yale sued the Berkshire Savings Bank for its use of a Consolidated time lock, Joseph Hall took a major step: he petitioned the court to have Consolidated substituted as the defendant in the case, contending that the maker was, in fact, the "real party in interest."[42] Under today's liberal rules of procedure, one rarely if ever sees a case where the party truly at the center of a controversy is forced to sit by and watch, but the legal landscape was much different in 1880. Two parallel court systems, courts of equity and courts of law, operated at every level, each with its own rules.[43] Patent infringement claims were considered claims in equity, which meant that while pleading and evidence rules were more relaxed, there was no right to try the case to a jury, only to the judge. But even with the more liberal equity rules, courts were reluctant to force plaintiffs to litigate against defendants that they had not sued, especially when they had a viable claim against their chosen foe. Prior to modern rules of procedure, deciding whom to sue was of paramount importance. A plaintiff's decision to sue two or more defendants at once (or "join" them) was reviewable at every level of appeal, and an error in joinder meant the case was dismissed completely. Adding insult to injury the statute of limitations was not tolled during this period. Hence, a plaintiff could find at the last appeal that the case was thrown out with no chance to refile. And so we find that motions such as the intervention by and substitution of a real party in interest were still cutting-edge legal ideas when Joseph Hall stepped into the suit against Berkshire Savings Bank.[44]

Not surprisingly, the trial court[45] reached the same decision as others, finding the Yale and Sargent reissued patents valid and Hall's design an infringing use.[46] But Hall was not one to take such a defeat lightly—he was, after all, the only maker to include his own nickel-plated portrait on every one of his flagship bank locks. Whether Joseph Hall thought that a damage award would put Consolidated Time Lock out of business or that he was truly right and would be vindicated by the courts, he decided to fight on, and in July of 1884 he received leave of the court to amend his answer and have the case reheard. Unfortunately, the rehearing court was no kinder to Hall, and on February 12, 1886, it affirmed the decree upholding the Yale and Sargent patent priority, setting damages owed by Hall at $60 per time lock and awarding a perpetual injunction against Hall, preventing him from producing his time lock designs.[47] Yet Hall persevered, and in 1889 he filed a petition for certiorari with the Supreme Court, requesting that the highest court in the land review his case. When the Supreme Court accepted the case, time lock litigation throughout the courts was thrown into disarray. The Supreme Court was unlikely to take a patent infringement case unless it felt it was affected by some substantive error of law, and many lower courts granted motions that their decisions be stayed, pending the Supreme Court decision.

In 1889, having stepped into the legal fray to defend his clients, having met with two resounding legal defeats and with the future of his Consolidated Time Lock Company in the hands of the court of last resort, Joseph L. Hall died. The Supreme Court noted simply that "his executors and trustees have been made parties in his place."[48] The substance of the

Major fires were an all-too-common occurrence in nineteenth-century urban centers and safe makers' reputations rose and fell on whether their safes' contents survived. The Park Row fire illustrated in this woodcut from a Marvin Safe Co. catalogue occurred in New York City around 1850.

Supreme Court's decision in *Berkshire* was nothing short of explosive, however. The Court found that Sargent's and Yale's original patents were not defective and that the sole object of seeking reissues was to obtain enlarged claims.[49] If these claims were patentable, the inventors should have made regular applications, not sought reissues.[50] Holding that the expanded claims of the reissued patents were void, the Supreme Court effectively reversed all lower court decisions in time lock cases dating back to the first decree of Judge Shipman in 1881, and vindicated not only the late Joseph Hall but also every time lock maker driven out of business by Yale and Sargent over the previous twelve years. For Hall himself and these other companies, the wheels of justice had turned too slowly. Makers such as Pillard, Lillie, Stewart, and Holmes had long disappeared from the market, never to return, but Hall's Consolidated Time Lock Company survived and would continue to be a major time lock maker, along with Yale and Sargent, for some time to come.

Some confusion has surrounded the origin, life, and ultimate fate of the Consolidated Time Lock Company.[51] Following Joseph Hall's death in 1889, both Hall's Safe & Lock and Consolidated Time Lock continued to function under the leadership of his sons. In May of 1892, Hall's Safe & Lock was sold, eventually becoming part of Herring-Hall-Marvin Co. of New Jersey, with Edward C. Hall, William H. Hall, and Charles O. Hall among the new company's shareholders. Herring-Hall-Marvin continued with Edward Hall as president and William Hall as treasurer until a dispute regarding employment contracts led all three Hall brothers to demand and receive releases from the company. In September of 1896, the three Hall brothers formed Hall's Safe Co., a company often confused with Hall's Safe & Lock, but in fact it was a completely distinct entity and not a successor company in any way.

In 1900 Herring-Hall-Marvin Co. (now without Hall-family participation for four years) was forced into receivership and emerged reorganized as the Herring-Hall-Marvin Safe Co. This new company conducted business for some time and then, in 1904, sued Hall's Safe Co. over its use of the Hall family name. This litigation lasted until 1906 when the Court of Appeals for the Sixth Circuit decided the case in favor of the Hall family, who were allowed to keep their name.[52] Hall's Safe Co. continued in business for more than twenty more years, finally closing its doors due to waning business in 1927.

Unlike Hall's Safe & Lock, Consolidated Time Lock Company was not sold by the Hall family so soon after the death of Joseph Hall. Consolidated continued making time locks under Hall family ownership from 1889 until at least 1906. Researchers will find an informational "dark age" in the early twentieth century regarding safe and lock makers due primarily to a dearth of litigation between the major companies. Inferences from production models suggest that Consolidated Time Lock ended operations around 1906. The company may have changed hands and had its name changed to Bankers Dustproof after being acquired by the Victor Safe & Lock Company. Bankers Dustproof was Victor Safe & Lock's wholly owned time lock subsidiary, first appearing in 1906 with time lock movements substantially similar to those of Consolidated. Victor Safe & Lock would, itself, eventually become part of Mosler.

With the advent of widespread reliable electrification at the end of the nineteenth century, bank security began to include electric alarm systems for the first time. The first public tests of electric alarms were met with skepticism—and, on occasion, some distress, as was the case in a New York exposition by Edwin Holmes. Holmes was an important pioneer, understanding the significance of electrical technology to the security business. After he bought the patent rights to an electric alarm system from Augustus Pope, he founded the Holmes Burglar Alarm Co. in 1857. Moving his company and family to New York City in 1859, Holmes was quite a success selling burglar alarms to both businesses and individuals, eventually inspiring the Mark Twain short story "The McWilliamses and the Burglar Alarm."[53]

Ultimately, Holmes's business focus would shift away from electric alarm systems. In 1872, Holmes secured a patent on electric sensors that were wired to and monitored at his central office in New York and a similar office opened in Boston employed Thomas A. Watson, who developed much of the Holmes company's best systems. After Watson left to work for Alexander Graham Bell in 1874, Holmes stayed in contact while he added the Holmes Electric Time Lock to his bank security business. The Holmes Electric Time Lock was a major success with many safe owners, and more than twenty years

The Hall family, who succeeded Joseph Lloyd Hall, circa 1893. Standing, from left: Walker Hall, Jess Hall Trevor, Sally Clark Hall, Harry Hall, Acton Hall, Pearl Hall Junkerman, and Charles Hall. Seated, from left: Anna Hall Pullen, Edward Clark Hall, Cloe Hall Kemper, Sara Jewel Hall, and Kate Hall Hart. A picture of Joseph Lloyd Hall is visible behind the family and one of Joseph Lloyd Hall, Jr., is to the right. (Courtesy of the Hall family.)

later, after Holmes's patent protection had expired, his basic design would be revived by the American Bank Protective Co. for use in its own bank alarm timers.

Holmes's New Electric Time Lock was widely acknowledged as superior and was adopted by a number of prestigious clients, including the United States Treasury Department. However, it was the use of a Holmes time lock by movement maker E. Howard that was seen as a major embarrassment by Yale, and beginning in 1878 safe owners using Holmes's time locks became targets of Yale's and Sargent's combined lawsuits. Holmes's clients lost or settled these suits, but Holmes himself was not easily dissuaded. He returned to the Patent and Trademark Office, securing his own expanded patent reissues,[54] and in 1879 he filed lawsuits against the Catskill National Bank for its use of a Beard & Bro. time lock[55] and, in a brazen counterstrike, against the Connecticut Mutual Life Insurance Company for its use of a Yale Pin Dial time lock.[56] However, it soon became clear that Holmes's time lock business was not sustainable. He did not have the patent portfolio to take on the Yale–Sargent combine and, unlike Joseph Hall after him, Edwin Holmes seems to have accepted that the expanded reissue of patents was a legitimate tactic. Besides, greener pastures were opening up for him in other areas.

Holmes's burglar alarm company was flourishing in New York and would continue well into the twentieth century but it was his relationship with Thomas Watson that yielded fruit in 1878 when Gardnier Hubbard, one of Bell's backers, asked Holmes to organize the Bell Telephone Company. Initially,

> # THE HOLMES ELECTRIC PROTECTIVE COMPANY
>
> **1824 — 1924**
>
> IN 1858 Mr. Edwin Holmes came to New York from Boston with the first electric burglar alarm. He found that the majority of people had little faith in his claims for this was years before the telephone had been invented. Electrical devices were not generally known. So he had constructed the model house shown here. This was small enough to be conveniently carried about. The door and windows were electrically connected to a battery inside and to the bell on the roof. Opening the door or raising a window caused the bell to ring automatically. Even with this he was often accused of practicing "black magic."
>
> In 1872 the Holmes Burglar Alarm Telegraph Company was organized to furnish service from Central Stations. In 1882 the name was changed to the Holmes Electric Protective Company. On January 1st of that year the Company had two Central Stations; one at 194 Broadway, the other at 518 Broadway; and served a total of 471 subscribers.
>
> To-day the Holmes System protects nearly three thousand banks, stores and residences in the Fifth Avenue section alone and serves Greater New York through fourteen Central Stations.
>
> *Model of* FIRST ELECTRIC BURGLAR ALARM
>
> *General Offices* · 370 SEVENTH AVENUE

An advertisement for the Holmes Electric Protective Company from the early twentieth century recalls a time when people were less familiar with the nature of electric mechanisms.

Holmes, Bell's first president, faced significant patent litigation against Western Union, which was supported by Thomas Edison, but this eventually ended when Bell Telephone bought the Western Union system in 1880. Shortly thereafter, Holmes sold for $100,000 his interest in Bell Telephone of New York, the company that would go on to become AT&T.[57]

Within ten years, electric bank alarm systems were well accepted and the business of supplying the security companies with timers for their systems began to expand. Most alarm system makers chose to rely on movements similar to those used in time locks, due to both their time-proven reliability and bankers' comfortable familiarity with them. These alarm timers are sometimes confused with time locks but have a number of identifiable differences. The most common and obvious difference is that alarm timers universally have a set of electric contacts, most commonly on the case top. With the two exceptions of Holmes's Electric time locks (which used an electric pendulum as a backup mechanism) and the Hollar time lock (which featured an electric rewinding system), time locks do not need these electric contacts. While the cases of time locks produced prior to 1940 are universally metal, the cases of alarm timers were often made of an insulating material, such as wood or Bakelite.

Early alarm timers relied on a single movement, while even early time locks featured two or more movements, a backup system (such as Hall's Infallible mechanism), or both (as in the Holmes Electric). This seemingly relaxed attitude toward the possibility of the alarm timer's failure was not entirely misplaced. In the event a time lock on a safe or vault failed, the contents would be completely unretrievable until the door was drilled—an expensive, destructive, and time-consuming undertaking. However, should an alarm timer fail to disable the bank's alarm at the set time, this would cause only a false alarm. With little or no cost associated with the alarm timer's failure, a single movement was initially sufficient, although later alarm systems did employ multiple-timer formats.

Concurrently with the rapid evolution of bank locks and time locks, safe and vault construction was advancing as well. What were then advanced designs for burglarproof vaults were newsworthy developments and were often the subject of media coverage. In 1894, *Scientific American* reported on the use of alternating layers of hard and soft steel to create the four-hundred-ton National Safe Deposit Vault.[58] This design effectively foreclosed the option of drilling into the vault, since with their different densities these layers would bind and break drill bits. Three time locks guarded each of the two solid doors, ensuring that the vault would open at, and only at, the appointed time. The later use of "Harvey-ized" nickel-steel armor plate in vault construction was another notable news item.[59] Built by the Hollar Company for the Provident Life and Trust Company, the vault was a two-story giant whose upper door alone weighed sixteen tons. Vaults such as these, combined with manipulation-proof combination locks and modular time locks, were all but unassailable and came to serve as a metaphor for a bank's security and stability.

CHAPTER 2

The Early Prominence of Key Locks: 1834–1856

The first half of the nineteenth century saw the nascent American banking industry rely primarily on the key lock to secure safes and strongboxes. More often than not these locks were based on the lever designs of Barron and Chubb, adding complications and narrowing tolerances. The industry was small, and although most important models were developed by a handful of established locksmiths, a few individuals saw some success making a single patented design. With machining itself still in its infancy, early lock manufacturers worked mostly by hand, filing and finishing their locks piece by piece, making these early designs even more wonderful. Though bank lock production would later develop into a major industry dominated by large corporations, much of their success would depend on these early makers to demonstrate the viability of a market for high security bank locks.

C. J. GAYLER'S BANK LOCK

The first American bank lock to offer security on par with its English counterparts (exactly so, in fact) was manufactured at least as early as 1834 by C. J. Gayler [1] of 102 Water Street, New York City. This lock is an American-made four-lever detector-catch version of the Chubb mechanism that Gayler began manufacturing as soon as Chubb's patent protection expired. While some references have suggested that the Gayler lock dates only to 1860 and was intended for fireproof safes,[1] this is not the case. While Chubb's patent protection in Britain may have lasted until as late as 1835, concurrent publications clearly place the debut of this lock at least as early as 1834.[3] And while a four-lever mechanism such as Gayler's would, in fact, have been relegated to fireproof duty by 1860, in the mid-1830s, this multiple-lever lock with its detector catch was state-of-the-art burglarproofing. As an American manufacture of a well-known English design, Gayler's lock was not innovative, but this lock is still important, for it marks the genesis of major safe lock manufacture in the United States. Gayler is thought to have made many of these locks, but fewer than five examples of Gayler's lock survive today.

Gayler's lock and key were faithful reproductions of Chubb's 1818 design and found similar success in the United States. The false key detector is clearly visible above the stack of six tumblers.

SOLOMON ANDREWS'S BANK LOCK

If the first secure American-made bank lock belongs to Gayler, the first commercially successful American-designed vault lock belongs to Solomon Andrews. By the mid-1830s, lock makers were well aware that a conscientious thief could eventually open even state-of-the-art locks, but they also knew that a picker who had seen the true key—even just once—had an enormous advantage. In 1836 Solomon Andrews patented a bank lock that could take familiarity with the true key off the list of advantages for the safecracker. Andrews called his key lock a "combination" lock, referring to the user's need to know how the removable bits should be assembled on the key. Andrews's key design featured a square cross-sectioned shaft onto which the user could slide the bits and a nut that would be threaded on afterward to secure them. The first and last bits were required in their positions, but any bit in between could be rearranged or substituted by a bitless spacer, allowing the key to be carried unassembled or misassembled.

Even if a true, assembled Andrews key was lost, the owner could simply change the keying of the lock, rendering the lost key substantially useless. One method of quick key change was the introduction of spacers in place of bits on the Andrews key. With the lock open, bitless spacers could be placed on the key shaft in place of any bit or bits, and once the lock was closed with this new spaced key the lock required this spaced key to open. The Andrews mechanism also allowed the owner to open the lock with a compromised key, change the bit permutation, and lock the lock with the newly ordered key.[4] This ability to quickly and easily change the keying of a lock was a significant advance in both security and economy, allowing managers to change which employees had access to a vault without concern about the number of people familiar with the true key, or to rekey the vault in the event an employee left or was fired.

Around 1842, Andrews introduced a seventeen-lever model of this design, which incorporated a Chubb-style detector

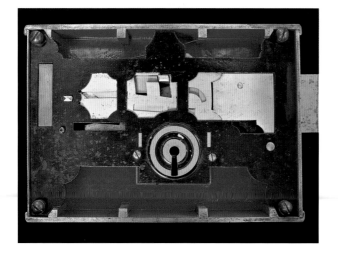

LEFT: *The interior view of Andrews's lock shows the thirteen levers and their arched secondary levers that separated the primary levers from the keyhole.*

OPPOSITE, TOP: *Andrews does not seem to have had a specific standard design of his lock. Seventeen levers and a Chubb-style false key detector made this Solomon Andrews lock one of his most secure. (Courtesy of Lynn Collins.)*

OPPOSITE, BOTTOM: *The key is seen here with four of its seventeen bits replaced by spacers. The four replaced bits show how each was embossed with a number corresponding to a specific lever. An open Andrews lock could be closed with a key with spacers in place of certain bits. Only this properly spaced key could open the lock, making spacing a quick and convenient way to change the key without opening the lock's case. Two additional spacers are shown. (Courtesy of Lynn Collins.)*

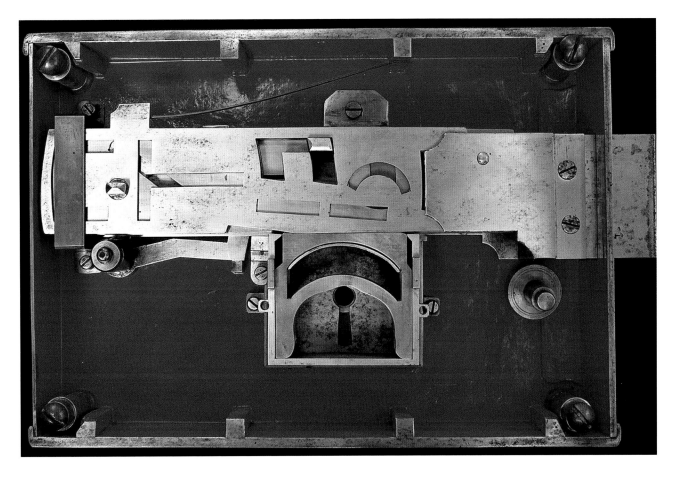

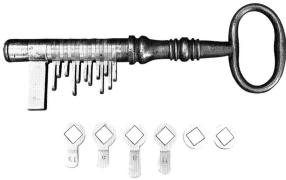

that could alert the owner to certain attempts to open the lock. Andrews's detector did not catch and hold levers overraised by a false key like Chubb's, but rather raised a flag visible from the interior when a lever that should be left idle by a bitless spacer was raised. This highly developed example of Andrews's important design features levers of alternating brass and iron to counteract the tendency of thin, light levers of like metal to stick together. Although this lock was removed from the Tanners Bank of Catskill, New York, around 1875, certain decorative cutaway sections of the case suggest that it may have been an exhibition piece as well.

There does not seem to be any organized evolution in the size or length of keys or the number of levers in Andrews's locks over time. Rather, they all seem to be one-off custom-made locks based on a common design.[5] However, an overview of many Solomon Andrews locks shows two basic styles: earlier models, with a key that operates directly on flat-bottomed levers; and later models with a key that operates on a set of vertically constrained intermediary plates with formed bellies, allowing for narrower, more secure gates and better disguising of which levers were "dead," corresponding to spacer bits.[6] No records of the total number of locks made by Solomon Andrews are known today, but ten examples are known to survive.[7]

OPPOSITE, TOP: *Solomon Andrews's bank lock marked the first time that an American-made design offered leading security with reorderable levers that meant the user could make basic changes to the key. For the first time, losing a key was not an expensive or risky misfortune.*

OPPOSITE, BOTTOM: *The interior view of this massive lock shows its seventeen primary and secondary levers. (Courtesy of Lynn Collins.)*

ABOVE: *Solomon Andrews's changeable bit key with bits that could be replaced by spacers. The key-changing process could be done quickly, by introducing spacers, or completely, by reordering the levers in the lock and the corresponding bits on the key. If the user chose to replace a bit with a spacer, the lock could be rekeyed simply by locking it with the new, spaced key.*

ROBERT NEWELL'S PATENT

Patent models of locks are generally very uncommon and a patent model from a well-known lock maker is quite rare indeed. Many of the early lock patent models were made of perishable wood, and even many metal models did not survive the fires that struck the United States Patent Office storage facilities from time to time. This lock was made by Robert Newell and is thought to be the patent model for this design. The main bolts that extend vertically from the top and bottom of the case (visible at the left) are thrown by a wrench attached to the larger socket and a secondary bolt can be extended from the case side by a wrench in the smaller socket. Much like a very heavy door lock, Newell's lock can be operated from the inside if the key bits are removed and reversed.

Generally, patent drawings emphasize a lock's improvements rather than closely corresponding to the final production design. In contrast, this lock corresponds closely to the schematics filed with Newell's patent number 944 from September 25, 1838. Another consideration is this lock's provenance, it having come through a reputable collector who obtained it from a large collection of known patent models. This patent model is the only known surviving example of this Newell design.

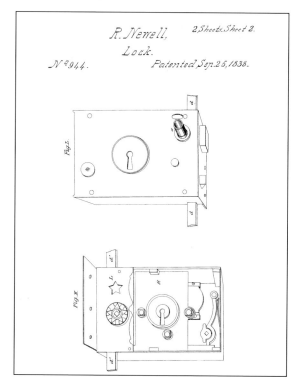

ABOVE, LEFT AND RIGHT: *This Newell lock was made unusually close to the patent specifications, seen here. Even certain decorative parts, such as the star-shaped bushing, are included in the patent drawings.*

OPPOSITE: *A Robert Newell patent model from 1838. Robert Newell was one of the most significant lock designers prior to 1850. This early Newell lock could be opened from inside like a door that once the key's changeable bits were reversed on the shaft. The interior view shows vertical bolt extensions that could be retracted with a second handle once the horizontal bolt had been thrown with the key.*

NEWELL'S PATENT

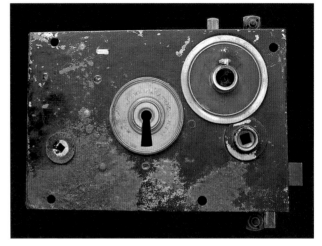

SOLOMON ANDREWS'S SNAIL WHEEL LOCK

The success of Andrews's changeable-bit key locks was not his only notable contribution. As early as 1842, Andrews's firm, Andrews & Maurice of Perth Amboy, New Jersey, was selling what would come to be known as his Snail Wheel Lock for $10 to $15. The name referred to the logarithmic spiral-shaped cutouts in the rotating discs that form the locking mechanism.[8] Each disc tumbler also has a notched edge to act as a gate for the fence, in this lock also referred to as the "toggle." As the key turns, the different parts of the bit engage each disc[9] as the narrowing snail hole is met by the bit with it. This allows each disc to be turned a unique radial distance, aligning the notches and admitting the fence.

While the basic principles of Andrews's snail lock were recent developments at the time, Andrews himself was not the original inventor. As part of the Great Exhibition of 1851 in London, an American named Jennings reportedly showed a lock with a changeable key that operated a stack of discs similar to that in Andrews's Snail Wheel Lock. Due to the design's lack of tumbler springs, Jennings billed his lock as both the least expensive and the best, his springless discs avoiding Hobbs's tentative picking method that opened the Chubb lock.[10] We have no record of Andrews raising any particular objection to Jennings's version of the snail lock. However, we do know that Hobbs, upon hearing of Jennings's boast, was able to quickly pick Jennings's lock with a piece of watch spring, using a hooked end to feel for and align the discs' notches, and that by 1857 Mr. Jennings and his lock were nowhere to be found in England.[11]

Andrews's springless disc design incorporated sufficient improvements over Jennings's model to make the Snail Wheel Lock a bank lock of significance. Beyond the "snail" disc design, he added a nonlocking front disc with only a keyhole and gate, to offer a first line of defense against probing picks. But what truly made Andrews's Snail Wheel Lock a formidable lock was the sheer size and mass of his discs combined with great pressure from the bar across the wheel pack and friction from the springs below them. While the tight assembly made Andrews's key unusually large and unwieldy even for vault locks, it also made it nearly impossible for a picker to ascertain the slight difference in radial turning resistance when the toggle is resting on an edge or over a notch. Further, the Hobbs watch-spring picking method was rendered useless since any probe sufficiently thin and flexible to reach all discs was almost certainly too weak to effectively manipulate them. The Snail Wheel Lock in the Mossman Collection was a gift of the Chemical National Bank of New York City, where it is believed to have been used in a vault until its replacement around 1900. This is the only known surviving example.

SNAIL WHEEL LOCK

Solomon Andrews's Snail Wheel Lock and detail of its keyhole. The Snail Wheel Lock relied on a short brass band to create the high compression that made Andrews's version far more secure than Jennings's original model. The wheel pack band is visible here toward the left of the case, secured across the middle of the round tumblers by two screws. For its price, the Snail Wheel Lock may have been the best security value of its time, but its heavy mechanism meant that users would have to deal with cumbersome keys and difficult unlocking.

LINUS YALE, SR.'S QUADRUPLEX LOCK

Linus Yale, Sr., is among the pantheon of great American lock designers. From the beginning of his lock making career around 1840,[12] Yale Sr. seemed to understand the growing demand for high-security bank locks and devoted both his time and his fortune to this still-nascent industry. In 1844 Yale Sr. patented the first version of his innovative Quadruplex key lock,[13] the first true modern adaptation of the 3,500-year-old Egyptian pin tumbler mechanism. This model and a smaller, second version for use on safes would remain among his most effective and successful lock designs. In 1849 Yale patented the larger Double Quadruplex lock.[14] Like Nock's vault lock from the 1820s, Yale's Double Quadruplex uses a secondary lock to guard the keyhole of the primary lock[15] but, unlike earlier double-custody designs, Yale's model also used two formidable locks of the same superior security for both primary and secondary functions.

Each Quadruplex mechanism has two pairs of horizontally opposed five-pin tumblers radiating from the keyhole at 45°, 135°, 225°, and 315° and the key's cylindrical bit has four fluted grooves corresponding to the tumblers. The Double Quadruplex lock's operation is somewhat of a puzzle and certainly added to its security. First the two-handled wrench is placed in the left lock and the smaller handle is turned 180° clockwise, retracting the plate that guards that left keyhole. The wrench is removed (already so, as shown) and the smaller, flat-topped key is placed in the left keyhole and its stem is unscrewed, leaving the bit inside. The two-handled wrench is replaced in the left keyhole, the small handle is turned 180° counterclockwise, the large handle is turned counterclockwise to a stop, and the small handle is turned a further 90° counterclockwise. This last turn retracts the hardened steel plate visible between the two locks, revealing the right keyhole. Finally, the large bowed key is inserted in the right lock and turned 90° counterclockwise, throwing the bolt open.

Yale is thought to have made a few hundred Double Quadruplex locks, of which five are known to survive today. Yale also produced a Single Quadruplex in a number of sizes, some quite small. Many examples of these smaller Single Quadruplex locks survive. Interestingly, the pins in the Quadruplex design operate much in the same way that our standard door locks do today, although modern keys have been greatly simplified.

Linus Yale, Sr.'s formidable Double Quadruplex bank lock. With bank lock design in its infancy, many early bank locks relied in part on a complex opening procedure for their security. With two locks, each with four sets of pin tumblers, the Double Quadruplex mechanism itself offered top security for its day. However, it also required a user to be familiar with its complicated operation, which included two bits and a two-handled key.

QUADRUPLEX LOCK

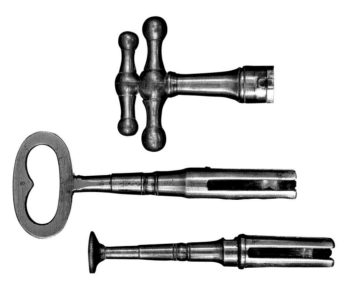

GEORGE WILDER'S SALAMANDER SAFE LOCK

With both burglarproofing and fireproofing of safes in their infancy, the earliest safe makers made little distinction between the two types of protection. But manufacturers soon realized that the methods of fireproofing were not well suited to burglarproofing. Fireproofing generally requires the safe maker to incorporate one or more layers of low-density substances between layers of metal to slow the conduction of heat from the outside of the safe to the paper contents. Common fireproofing layers used substances such as clay, plaster, or cement.[16] Some makers simply left air spaces between metal layers. As early as 1845, fireproofing layers had become complicated mixtures, such as that used by James Watson & Son, Philadelphia, composed of "plaster of Paris, asbestos, soapstone, and ising-glass, or some one or more of these substances combined."[17] However, such layers are simple for safecrackers to drill, cut, or chip through and weaken the metal layers above them, compromising the safe's burglar-resistance.

Burglarproofing requires materials and designs that hinder drilling or cutting and can withstand explosive force, limiting makers to steel, forged iron, and other tough metals. While layering these metals with varying hardness and softness makes drilling or cutting a slow and laborious process, these tightly bonded layers of highly conductive metal do little to protect the contents from heat. Manufacturers found that a safe of a given volume could be made burglar-resistant or fire-resistant but an attempt to combine both qualities yields a product that is neither particularly fire- or burglar-resistant.[18] By the mid-1800s, safes were advertised as either fire- or burglarproof, but generally not both.[19]

There is some evidence that a patent was sought for a specifically fireproof safe as early as 1830 by a plaster image maker named Fitzgerald. This patent seems to have been refused and the idea sold to Enos Wilder. Enos Wilder had no better fortune seeking the patent and transferred the idea to his brother George.[20] Not until New York's great fire of 1835 tore through hundreds of buildings, bankrupted twenty-three of New York's twenty-six insurance companies, and showed standard burglarproof safes of the day to offer little protection for their paper contents was the U.S. Patent Office willing to entertain fire-resistant layers as an improvement in safes.

In 1843 George Wilder of Brooklyn, New York, secured his patent and safe makers Rich & Co. began to offer Wilder's Salamander safe. The "salamander" designation was part of a tradition of labeling fireproofed safes as "salamander" models, drawn from folklore that held the salamander to be impervious to fire. Asbestos was sometimes referred to as "salamander wool" and references to salamanders' immunity to fire can be found in the works of Aristotle and Leonardo da Vinci and in sources as diverse as Egyptian hieroglyphs, medieval European manuscripts, and thirteenth-century Chinese writings.[21] Used more often than any other name by many companies throughout the 1800s for fireproof safes, the salamander designation was the source of any number of advertising disputes over more than thirty years among safe makers claiming priority to use this popular name. Consequently, the salamander designation did not denote any specific type of construction but rather a safe maker's top fireproof design.

SALAMANDER SAFE LOCK

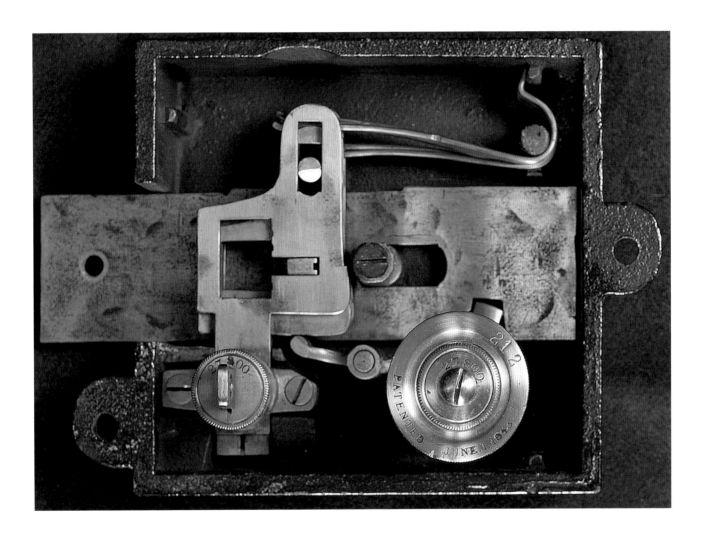

OPPOSITE: *The salamander's mythic resistance to fire made it a popular image among early fireproof safe makers. It is not clear who first used this creature on a fireproof safe, but it became so well known that eventually safe makers simply used the term "salamander" to indicate their premier fireproof model.*

ABOVE: *Burglarproof safes for banks and large businesses commonly protected contents worth far more than any safe on the market, which meant that the safe maker could justify the most expensive of locks. Safes like Wilder's Salamander, however, were designed to fend off heat, keep honest people honest, and be affordable for individuals and small businesses. Consequently, fireproof safe makers sought a balance of security and expense in their locks.*

The lock used in Wilder's Salamander safe is a simple five-lever key lock with a trunnion that is turned to lift the tumblers onto the inserted flat key and retract the bolt in one motion. Wilder's use of this lock made clear that the purpose of this safe was fireproofing rather than burglarproofing since a diligent picker could open this lock with little complication. This aside, Wilder's Salamander was very popular among individuals and small businesses and production continued as late as 1870. Similar safes were offered by Lillie, Terwilliger, and Marvin using similar locks. These locks are not particularly rare even today, and though they represent an important niche in American safe making history (in part due to their mundane construction) they were rarely preserved beyond the life of the safe. The Wilder Salamander Safe lock from the Mossman Collection is complete and in exceptionally fine condition. Wilder's lock is of interest in the context of safe and vault development, yet most notable lock innovations were made with burglarproofing in mind.

OPPOSITE: *With eighty-three moving parts, Robert Newell's Parautoptic lock may be the most complicated key lock ever designed. The Parautoptic was the culmination of Newell's efforts to develop the unpickable lock by isolating the bolt and the tumblers that act on it from the keyhole. Newell used three sets of ten tumblers, one that interacted with the key, one that interacted with the bolt, and a third set of intermediary tumblers that all but eliminated the ability to determine the position of the bolt tumblers through the keyhole.*

DAY & NEWELL'S PARAUTOPTIC LOCK

Safe lock designers of the 1850s were still in search of the truly "unpickable" key lock. Robert Newell, a partner at the firm of Day & Newell, focused on how the picker worked, exploring the inner workings of the lock through the keyhole. Newell sought to prevent this exploration and in 1844 patented[22] what Day & Newell originally marketed as the New American Permutating Lock,[23] but better known today as the "Parautoptic" lock. The term *parautoptic* was coined from the Greek, meaning "concealed from view," referring to the difficulty of observing the internal workings through the keyhole. The Parautoptic design included innovations Newell had patented in 1838 and 1843 and was certainly the most complicated lock of its time and one of the most complicated ever. It employs three sets of ten tumblers, one of which moves with the bolt, and twenty-two springs among the eighty-three pieces in motion when the key is turned, all contained within a twenty-one-piece case and a four-piece cover.[24]

One of the Parautoptic's major advancements was the method for rekeying the lock. Previous designs such as Andrews's lock relied on the user to introduce spacers in place of key bits for quick changes or to open the lock itself and reorder the internal levers to match the newly ordered key bits. By contrast, the Parautoptic allowed the user simply to

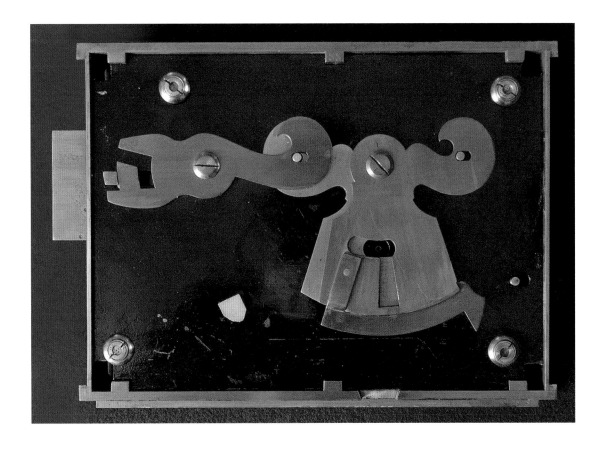

The Yale Lock.

Editors of the Boston Post.

No. 248 Front St., Philadelphia.

Gentlemen:—The article under the caption of "A greater than Hobbs," which appeared in the columns of one of your late issues, is, we believe, likely to be the involuntary means of doing us harm, from the similarity of the names of the patentees of the duplex lock therein spoken of, and of those which we ourselves manufacture; we would, therefore, ask room in your sheet for the following explanation, which, it seems to us, the more necessary to make, as our friends have forwarded us the article above spoken of, which they have clipped from your and other papers which have copied it, in order that we might defend ourselves, if unjustly attacked, they supposing, (from the similarity of the name) that ours is the lock spoken of as the "Yale Patent."

The Duplex lock (Yale's patent) is the invention of the late Mr Linus Yale, Sr., who sold the patent right to Mr Bacon, the present manufacturer of a variety of locks under it. These locks are entirely different, in every respect, from the locks of our manufacture, and known as the "Infallible," the "Magic," and the "Treasury" locks, which are made under the various patents of our Mr Linus Yale, Jr. We may add, with just pride, that ours not only have not been picked, but, also, that our challenge of $3000 is now, as formerly, open to the public.

Do not understand us to doubt the statement in the article referred to, that the Duplex locks can be picked, for our own manipulations on it have long since demonstrated the ease with which it can be accomplished, and it is now some years since we published to the world the weakness of Mr Bacon's locks in this respect; and we are equally well aware that the "Parautoptic" or great "Hobbs Lock" has been picked, since our Mr L. Yale, Jr. was indisputably the first to accomplish the feat as long ago as the year 1855.

Yours respectfully,

Linus Yale, Jr., & Co.

The Parautoptic lock was considered unpickable for more than ten years. Linus Yale, Jr.'s ability to open it brought him national recognition, the fact of which he would remind the public often.

open the lock, reorder the key bits with or without spacers, and then close the lock with this new key. The Parautoptic's tumblers would adjust themselves, granting the user easy access to the entire range of bit combinations.

For all its complications, even the Parautoptic lock eventually proved vulnerable to someone with sufficient time and patience. By 1855 Linus Yale, Jr., could pick the Parautoptic lock, which he publicly asserted in a letter published by the *Boston Post*.[25] Further, Yale patented the minor improvement that would have fixed the weakness he exploited,[26] precluding Newell from using it. Yale then traveled the country demonstrating his ability to pick the Parautoptic lock, and by 1860, unable to recover from Yale's attack, Day & Newell seems to have stopped production of locks entirely. The Parautoptic with Yale's improvement was made for at least some time in England by the London firm originally founded by Hobbs. Whether this activity was beyond the scope of Yale's patents or beyond his interest to pursue is unknown.

It is also unclear how many locks Day & Newell made. Based on the relative abundance of known locks and keys there may have been even fewer than those made by Solomon Andrews. While there seems to be similar numbers of surviving Solomon Andrews and Day & Newell locks, more Solomon Andrews keys are extant today. Fewer than twenty-five examples of the Parautoptic lock exist today.

opposite: *The Parautoptic's key featured ten changeable bits that could be reordered prior to locking. For the first time, the user could rekey the lock without opening the case.*

PARAUTOPTIC LOCK

D. M. SMITH'S BANK LOCK

In 1846, D. M. Smith of Springfield, Vermont, patented a bank lock[27] that at first glance seems to be a throwback, even for its early place in time, using the same wrench operation found in the Perkins and Rodgers locks. But without multipart pins to block the bolt, Smith's design was not among the precursors to later pin tumbler locks in the way the Perkins or Rodgers designs were. Instead, Smith included a fence balanced between two leaf springs: one below that tends to raise the fence, disengaging it from the teeth on the bolt, and a second above that can drive the fence down onto the bolt's teeth or allow it to rise away from them depending on the position of the eight wrench-driven pins.

Although it received no fanfare and little acceptance, Smith's design was the harbinger of things to come in combination locks. His fence mechanism is a nascent form of the drop lever that would eventually become standard in bank combination lock mechanisms. These later models invert Smith's design and rely on gravity rather than on a leaf spring to move the fence toward the tumblers, freeing the boltwork. Interestingly, what seemed to be a simplifying replacement of a spring with gravity would soon be deemed a weakness, since a gravity-operated drop lever tends to rest on the tumblers, allowing safecrackers to judge the tumblers' positions based on their interaction with the drop lever.

Smith's design, like J. H. Butterworth's, also included numbered dials for reference when turning the spindle, Smith's being a wrench with a brass collar numbered 1 through 9. The example of Smith's vault lock in the Mossman Collection was found under a freestanding vault in the Mutual National Bank in Troy, New York, suggesting that this lock may once have been installed there. Only one other example of D. M. Smith's vault lock is known to exist.

D. M. Smith's bank lock featured an eight-number combination that the user entered one number at a time on each of eight arbors. A barbed fence floats above the bolt, balanced between two leaf springs. The first true combination locks allowed this fence to rest on the tumblers, relying on gravity to pull it down. It was a simplifying change, but it would soon be identified as a security weakness. Improvements in machining accuracy and additional mechanisms such as James Sargent's Magnetic and Automatic designs of the mid-1860s overcame this flaw, one that Smith's lock never suffered from.

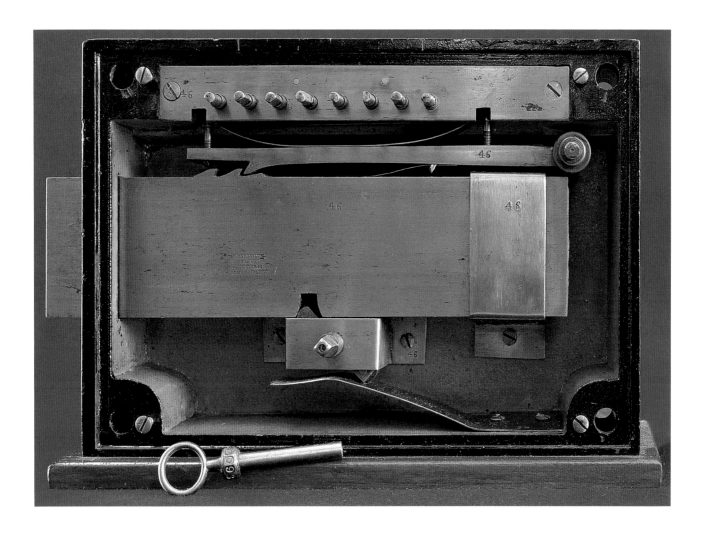

J. H. BUTTERWORTH'S COMBINATION LOCK

Also patented in 1846 was J. H. Butterworth's combination lock. This lock uses four stacks of three to five tumblers, all of which must be properly set to their individual combinations before the bolt can be thrown. Like Smith's design, the tumblers are operated by a removable wrench with a numbered collar, which seems to be a concession to the then prevalent expectation that a lock would have a removable "key." Butterworth's lock features a back cover locked by what he called a "dial lock," with another four-tumbler combination mechanism and the now familiar fixed dial. This is the first significant lock to incorporate a feature to prevent unauthorized access to the interior of the lock, possibly an accommodation to evolution in the banking industry away from close-knit family- and partnership-run operations to corporate structures with more employees and a greater division of labor.

In sharp contrast to a painfully complicated mechanism like the Parautoptic lock, Butterworth's lock is almost a picture of simplicity: four tumbler stacks guard the fence, yielding fewer than twenty-five moving parts. But with a ten-number dial key working four three-tumbler stacks, this lock presented the thief exactly one trillion possible combinations. Prior to James Sargent's 1860 invention of the micrometer that could identify the minute change in the position of the fence as the dial was turned, such a vast number of possible combinations was truly a daunting obstacle. A model of this lock won a silver medal at the Twenty-fourth Fair of the American Institute in 1851 for excellence in safe locks.[28]

Despite prior patents for permutation lock designs,[29] Butterworth's lock is appropriately identified as the first of the dial-combination bank locks. No examples of the earlier locks survive, if they ever existed, and while the true scope of the banking industry's adoption of Butterworth's design is unknown, it was sufficiently well known to be mentioned by name as an option in Herring's 1855 production catalogue.[30] The excellent example of the Butterworth lock in the Mossman Collection was removed from a vault door at the Bank of New York, where it guarded the first safe deposit vault in the United States.[31] Five Butterworth combination locks of various designs are known today.

OPPOSITE: *The first modern combination lock was patented in 1846 by J. H. Butterworth. A single ten-number dial is used like a key to open each of the four three-tumbler combination locks. Relatively uncomplicated and easy to use, Butterworth's lock was among the most secure locks of its day and set the new direction that bank lock design would take in the United States.*

WILLIAM HALL'S GRASSHOPPER LOCK

William Hall of Boston, Massachusetts, was a prolific inventor of locks during the nineteenth century. Though he primarily focused on prison locks, he patented an unusual six-lever lock in 1848 that would go on to win a medal at the 1851 Great Exhibition as a gunpowder-proof lock.[32] Offering only modest security even for its time, this popular lock would become known as the "grasshopper lock" due to its operation and was commonly used in both strongboxes and fireproof safes.

The key is crescent-shaped, with six lateral pins evenly spaced along its length that correspond to the levers. The key is pressed into the heavily sprung keyhole and the handle is turned, moving the attached flange over the key and pushing the key into position to align the tumblers and allow the handle to turn completely, throwing back the bolt. When the door is closed, the handle is returned to the locked position and, as the opening in the flange exposes the key, the heavy lever springs eject the key with considerable force, hence its entomological moniker. The security weakness of the lock is that the levers are directly accessible through the keyhole, and with no complications to impede the manipulation of its tumblers this lock was formidable only to the honest. The particularly fine example of Hall's Grasshopper lock from the Mossman Collection was manufactured by the Herring Company of New York under license from William Hall.[33] While this single-bolt version was used on strongboxes, a larger three-bolt version was made for use on safe doors. Today, the grasshopper lock is among the most common locks to have survived from its era.

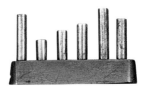

William Hall's Grasshopper lock uses its own heavy lever springs to eject the key after the bolt has been thrown. Its solid construction and small keyhole made it particularly resistant to being blown apart by gunpowder. Hall's Grasshopper lock was used extensively by the safe companies of Silas Herring.

GRASSHOPPER LOCK

E. M. HENDRICKSON'S BANK LOCK

E. M. Hendrickson was a safe and lock maker and his company, Hendrickson Safe Co., operated for more than twenty-five years at 67 North Oxford Street in Brooklyn from its beginning in 1855. Hendrickson specialized in opening competitors' safes by completely dismantling them from the outside, and by 1874 he claimed to have opened more than 760 different models built by such eminent makers as Herring, Farrell & Co., Valentine & Butler Safe & Lock Co., Marvin & Co., and Terwilliger & Co.[34] Hendrickson also invented a number of improved safe locks, but patented none of these.[35]

One example of a Hendrickson safe lock was a grasshopper-type design. Hendrickson's model offered somewhat greater security that the original, using a nine-lever mechanism in lieu of William Hall's six levers. However, this model truly surpassed Hall's original design with the addition of a changeable bit key, allowing over 360,000 key combinations. In contrast to Hall's model, this example of Hendrickson's burglarproof safe lock may be the only surviving model. In very fine condition but undated, the lock was likely made before 1860.

ABOVE: *Hendrickson's changeable-bit grasshopper-style key.*

OPPOSITE, TOP: *Hendrickson's grasshopper-style lock improved on W. Hall's design, with nine changeable levers. Ultimately, however, the grasshopper lock's exposed levers limited its security.*

OPPOSITE, BOTTOM: *Interior view of Hendrickson's lock showing the stack of levers and bolt.*

FRANCIS PYE'S TIME LOCK

Among the Mossman Collection is the time lock long identified as "Beard's time lock"[36] because the donor seems to have indicated that it was acquired from the firm of Beard & Bro. However, while Beard & Bro. began in the safe business in 1852 or 1853 and were safe, lock, and (later) time lock makers,[37] the first time lock that can be attributed to Beard with confidence does not appear until 1878, and even then Beard's designs were all patented by Phinneas King and show no relation to this piece. The construction of this time lock is important to note. Its single twenty-four-hour movement has two main spring barrels, ensuring even power to the large platform escapement with an unadjustable balance wheel, and the large, ivory-handled wrench winds both springs simultaneously. Overall, this time mechanism is of such high quality that very few clock makers could have produced it. While there is no known record of this movement being made by a major clock maker, the quality and style suggest the workmanship of Howard & Davis, a predecessor of E. Howard that made high-quality clock movements during the 1840s and '50s. Further, since this time lock's designer sought to avoid accidental lockout by using an extremely expensive, high-quality movement, it most likely predates any substantial testing of time locks. As early as Holbrook's 1857 design, it had become clear that a second average-quality movement was the most efficient anti-lockout solution. Unlike Beard's known time locks, there are no known contemporary images or descriptions of this one and the great expense, high quality, and attention to detail are far more consistent with an exhibition piece than a patent or production model. What can be said with some certainty is that, although this time lock may eventually have come through the hands of Beard & Bro., it was designed and made by someone else.

In 1851, the Twenty-fourth Fair of the American Institute was held at Castle Garden in New York City as a forum for inventors to showcase their new devices. Among the awards given by the jury were medals for bank locks, with the top-honor gold medals shared by a key lock by Newell and a more intriguing lock described in the judges' report as "a bank lock with a chronometer attachment" exhibited by Francis Pye.[38] Interestingly, mention of Pye's chronometer attachment can also be found in the defendants' answers of New Haven Savings Bank and Norwich National Bank in their patent infringement suits of 1879[39] and 1880,[40] respectively, referring to an 1854 article in the *Daily Morning Journal and Courier* and an 1867 article in the *True American*, both publications originating in New Haven, Connecticut. Unfortunately, despite all these references under oath to a chronometer-based bank lock by Pye, no known patent, illustration, or detailed description of Pye's mechanism survives.

Although Pye did not take out a patent on his time lock, he did patent a bank lock in 1846,[41] which is thought to be his award-winning design, although no examples of this lock are known to survive either. But this patent clearly places Pye

PYE'S TIME LOCK

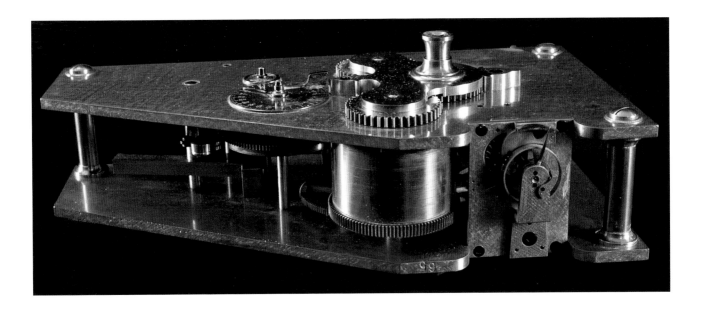

OPPOSITE, TOP: *Pye's time lock. Long assumed to have been made by the Beard & Bro. safe company, this time lock is now thought to be Pye's award-winning chronometer attachment from the 1851 Fair of the American Institute. Predating the earliest known commercial use of a time lock by seven years, Pye's design is beautifully constructed throughout, with a nickel-plated arbor shield and a twenty-four-hour dial made of silver.*

OPPOSITE, BOTTOM: *Pye's time lock has no mounting points on the back. The pins on the bottom suggest that it was used only for exhibition purposes.*

ABOVE: *Pye's lone platform escapement is powered by two barrel springs, one of which is visible in this top view. The elegant unadjustable balance wheel is mounted in a platform escapement and the bolt-blocking mechanism is visible on the left.*

The extravagant construction of Pye's time lock extended to its ivory-handled winding key.

among those inventors of the 1840s who were focused on bank vault security. Pye's "chronometer attachment" is the only time lock for which we have firm evidence, but no description. Further, this example is the only known time lock that is entirely unattributable to any known maker—it shares no clear markings, style, or components with any other known design. Based on these factors, combined with its construction placing it firmly in the period of Pye's work, this time lock is now thought to be the gold-medal chronometer attachment designed by Pye, the earliest surviving example of any time lock.

Prior to Pye's 1851 gold medal, patents were issued for time locks to Rutherford in England[42] and to Savage in 1847 in the United States.[43] Both of these patents were cited extensively in various patent suits throughout the 1870s and 1880s, but there is no indication that either was ever produced or sold. Pye does not seem to have sold his "chronometer attachment" either, suggesting that this model is unique.

OPPOSITE: *The Evans & Watson combination lock. Notched tumblers and a spring-loaded positioner made this 1852 Evans & Watson combination lock easy to use and inexpensive to make. Combination lock makers often had to contend with not only the limits of their own production accuracy but also the accuracy of their clients. Combination locks with many closely spaced numbers offered greater security but could tax the skill and patience of the day-to-day user.*

EVANS & WATSON'S BANK LOCK

Eventually, safe makers and lock makers would specialize, focusing on their particular areas of expertise, but during the early period of safe lock development, in the 1850s, it was still common for safe makers to offer locks of their own design. Evans & Watson was a safe making firm first established in 1838 by David Evans, adopting this name when James Watson joined in 1843. Originally fireproof safe makers, Evans & Watson began producing burglarproof models in 1850 at their locations at 53 South 40th Street and 248 North 8th Street, Philadelphia. Eventually, Evans would depart, leaving Watson to continue as Watson & Son.[44]

By at least 1852,[45] Evans & Watson was offering an unusual combination lock of its own design. The case is cast iron and the mechanism is brass. The nickel-plated dial, a removable knob with a thick, lettered spindle, clearly predates the low-profile, fixed, numbered dial standard by the 1860s. This dial is inserted and pushed to the back of the case, engaging the four tumblers that are turned discrete amounts with a clicking mechanism visible to the upper right of the tumbler stack. After the combination has been entered, the handle is pulled outward to engage the bolt at the bottom of the case, and then turned to throw it. Like many locks specific to safe companies of this time, few are thought to have been made and this example from the Mossman Collection may be the only one remaining.

H. C. JONES'S BANK LOCK

By the early 1850s, the lock maker H. C. Jones of Newark, New Jersey, was already well known in the lock world. Today, Jones's popularity survives based on the widespread enjoyment of the padlocks his company made, many featuring subtle and complicated tricks.[46] In his time, however, Jones was known primarily as an accomplished maker of safe locks,[47] such as his 1852 changeable-bit lever lock. It is based on a pair of patents, one granted to H. Ritchey,[48] thought to have been a Jones employee, and another granted to Jones himself.[49] Ritchey is known to have gone on to make padlocks under his own name.

A common feature of Jones's keys was the pin projecting from the end of the last, fixed bit. The changeable bits were held on with a set screw and the levers were set to accept a reordered bit when locked with the new key, much like the Day & Newell locks. Jones's locks were popular in their day and as a catalogue-listed option on S. C. Herring safes;[50] some H. C. Jones safe locks were likely made by Herring under license. However, as with many locks from this period, few Jones safe locks survive. Only two examples are known today, one of which was likely made by Herring.

ABOVE: *Reminiscent of the changeable-bit key of Newell's Parautoptic, Jones's six-bit key included a Stansbury-like pin on the end.*

OPPOSITE: *H. C. Jones's 1852 bank lock was a fairly simple six-lever design notable more for its maker than for its mechanism. Jones incorporated many well-known features, such as a changeable-bit key, but did not offer any significant advancements.*

H.C. JONES'S BANK LOCK

ALBERT BETTELEY'S SAFE LOCK

As early as 1841, Albert Betteley[51] started as a safe and lock maker in Boston, and by 1852 he had introduced his changeable-bit twelve-lever safe lock.[52] The large, finely made lock offered excellent security for its day, at the expense of somewhat cumbersome operation. First, the key is placed in the keyhole and the shaft is unscrewed from the bit, leaving the bit in place. Next the wrench is attached and turned, first obscuring the keyhole, then lifting the bit and pressing it onto the levers and, finally, throwing the bolt open.

While there were similar detachable-bit key designs made during the same period by the firm of Day, Newell & Miner as well as by Linus Yale Sr. and Jr., Betteley was awarded a patent for his design in 1852.[53] It is not clear at this time just who first introduced this design, but Betteley's detachable-shaft key design was quite effective and would appear in a number of locks by other makers.

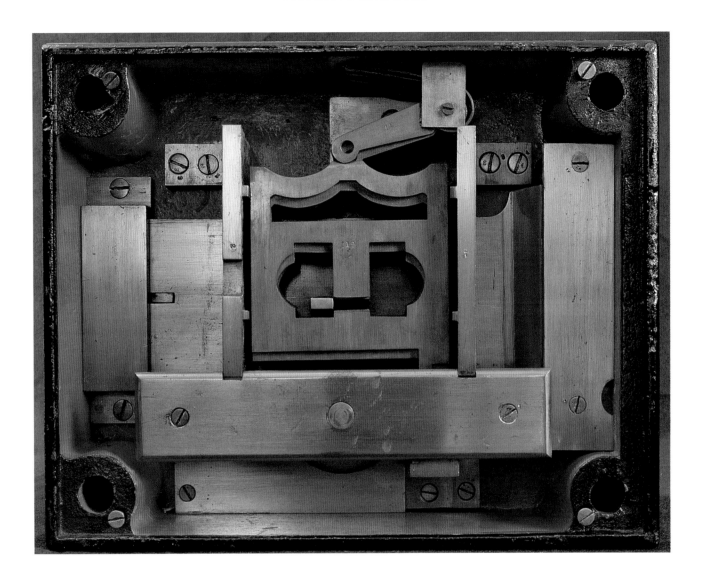

Albert Betteley's safe lock. Early lock makers strove to put levers out of reach of lockpickers, employing wards, keyhole blocks, and in Betteley's case, a two-part key. Multipart keys and complicated operation would eventually disappear as security features, but when Betteley's safe lock was introduced, it was as secure and user-friendly as any contemporary. The bowed unbitted shaft held the lever tension spring (visible above the lever stack) during rekeying.

LINUS YALE, JR.'S MAGIC KEY LOCK

In 1851, Linus Yale, Jr., introduced a new key[54] that used a series of discs attached around its shaft to create a three-dimensional bit. Soon improved to allow these discs to be rearranged,[55] these keys formed the basis for a series of three increasingly complicated versions of a bank lock that would become known as Yale's Magic Key Lock.

The earliest two types of the Magic Key Lock featured this round changeable bit attached to the key shaft by a finely machined dovetail. This allowed the bit to detach entirely from the shaft and travel away from the keyhole and up to the levers as the remaining shaft is turned. As the lock is closed, the bit returns and reattaches itself to the shaft. Although various forms of the keys for these two early styles of Magic Key Lock can occasionally be found, only three examples of the locks are known, two of which are seen here.

The third style of the Magic Key Lock involved a major revision to the key to accommodate a more sophisticated system of levers. With the number of levers increased from as few as four in earlier models to sixteen, the new key featured eight asymmetrical numbered bits, each operating a pair of levers with one bit on either side. One set of bits is seen here, with the other, differently keyed side obscured from view.

ABOVE: *The earliest version of Linus Yale, Jr.'s Magic Key Lock. The dovetailed separation point on the key is clearly visible to the left of the round bits. The spur on the key allows the shaft to act as a wrench by fitting into the notched collar around the keyhole after the bit has detached. When the key is turned, the pod of discs separates and travels into the lock, where it engages the levers.*

ABOVE: *The second version of the Magic Key Lock and the second version of the Magic Key.*

OPPOSITE: *Yale's last version of the Magic Key Lock added a second set of levers, increasing the total to sixteen, all of which were nearly inaccessible through the keyhole. It is seen here with the key spur collar and supporting bracket in place and removed.*

MAGIC KEY LOCK

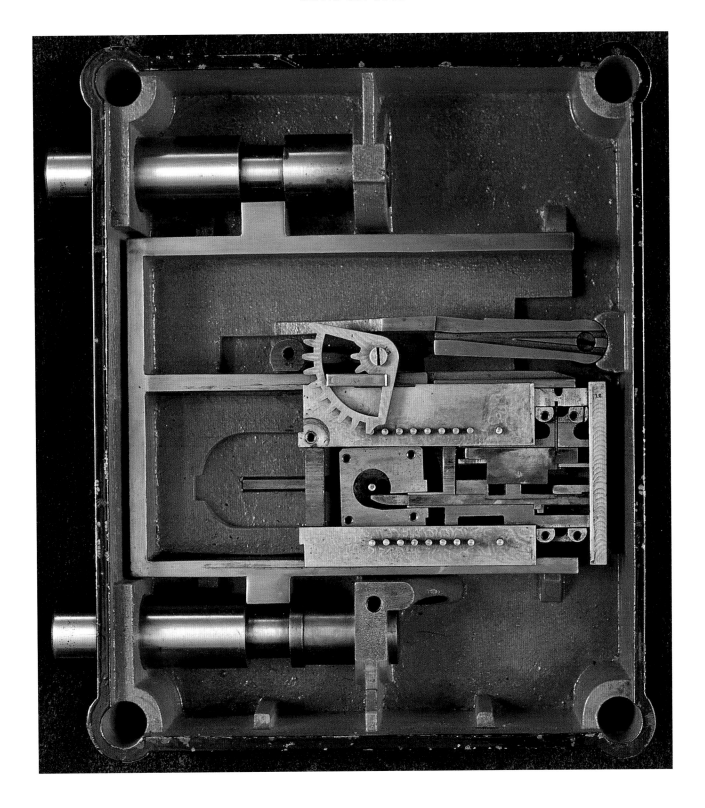

MAGIC KEY LOCK

This new detachable bit was held in place by a wedge-shaped collar and was released by a depressible pin that sat flush with the key's end. Once the key was inserted into the lock, Yale's new design worked much like earlier versions of the Magic Key Lock, with the bit detaching from the shaft and traveling through the mechanism as the shaft is turned.

Records do not survive of how many Magic Key Locks Yale made. Fewer than fifteen examples of these keys and only two examples of the lock itself are known. Considering the importance of Linus Yale, Jr.'s locks and the fine reputation that he always maintained, collectors have long found this relative dearth of surviving Magic Key Locks and keys to be something of a mystery.

The bits on the last Magic Key were no longer round, replaced by two sets of eight bits that sat flush against the faces of the detachable portion of the key. One set of bits is visible here. The other, asymmetrical set is on the reverse side. Like earlier versions of this key, the pod of bits travels into the lock and engages the levers as the key is turned.

ROBERT N. PATRICK'S PUSH-KEY LOCK

During the 1850s, there was a period of popularity of push-key locks that prompted many makers to introduce their own version of these relatively simple designs. These push-key lever locks were generally not patented and became the standard fireproof safe lock of this era. One push-key lock considered to be among the best for fireproof safes was an eight-tumbler model made by Robert N. Patrick.[56] Patrick manufactured fireproof safes at his facilities at 192 Pearl Street and 60–66 Cannon Street in New York between 1852 and 1870,[57] with his main model known as the Defiance Salamander safe. This lock, offering ease of use, low cost, and modest security was well suited for use in Patrick's Defiance Salamander.

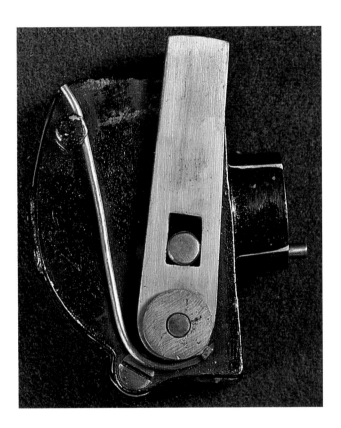

Like other locks intended for fireproof safes, Patrick's push-key lock could be inexpensively produced and easily used, but it was no match for a skilled burglar.

ABOVE, LEFT: *A side view of Patrick's compact lock design.*

ABOVE, RIGHT: *Patrick's key received only cursory finishing, since the mechanism did not demand the same close tolerances required by complicated burglarproof designs.*

LINUS YALE, SR.'S BACK-ACTION LOCK

After Alfred Hobbs picked the "undefeatable" Chubb and Bramah designs in 1851, lock makers began to realize that the ultimate weakness of any key lock was its keyhole. Although Linus Yale, Sr., never did away with the keyhole entirely, he did design the Back-Action lock, patented in 1853.[58]

With its detachable key bit, the Yale Back-Action offered top security for its day, due in part to its unusual operation, which alone could have confounded a potential thief. The Back-Action design put the true keyhole all but out of reach from the safe's exterior, placing it on the back of the lock's case. To open the lock the user attaches the handle and turns it 180 degrees, aligning the threaded rod on the armature inside the safe with a hole in the door. The handle is pulled out until the threaded rod extends through the door and the key bit is screwed on. The handle is pushed in, turned back 180 degrees, and pulled out, moving the bit up to and into the rearward-facing keyhole. With the bit in place, a second upper handle is turned to throw the bolt.

Ultimately, the use of black powder and nitroglycerine to unhinge safe doors would render the Back-Action lock obsolete, not due to the lock itself but rather to the opening through the door that the design required. Few Back-Action locks are thought to have been made and the example in the Mossman Collection is currently believed to be the only one to survive.

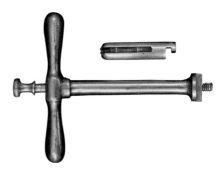

ABOVE: *The Back-Action lock's Quadruplex-like key bit does not require a turning motion, but rather aligns the pin tumblers upon insertion. This allowed Yale to position the keyhole completely inside the safe itself.*

LEFT: *The Back-Action lock of Linus Yale, Sr., shown here with its wrench and bit in place. The Back-Action lock itself was quite secure, but its weakness lay in the opening through which the bit was attached to the internal armature. The use of gunpowder and, eventually, nitroglycerin to open safes rendered many locks with large keyholes and through-door openings like the Back-Action lock obsolete.*

WILLIAM HALL'S CHANGEABLE BIT KEY LOCK

In 1854 William Hall, inventor of the still-popular Grasshopper lock, was awarded a patent[59] for a lock identified in *Lure of the Lock* as the Elward Safe Key Lock.[60] This attribution to an "Elward" is questionable, since no Elward has come to light anywhere else in safe lock literature. There is some speculation that this lock may be the work of Elwell, a lock maker in business at that time, but there is no firm evidence to support such an attribution. This lock would probably be more accurately described as William Hall's Changeable Bit key lock.

The lock is operated by inserting a shaftless bit into the keyhole and turning a wrench to throw the bolt. The shaftless bit is similar to Hall's own Grasshopper key, but this model offered a major improvement in security over the Grasshopper lock, with seven changeable levers that were difficult to manipulate through the keyhole.

LEFT AND OPPOSITE: *W. Hall's Changeable Bit lock worked similarly to his popular Grasshopper lock but placed the levers to the side of the keyhole rather than directly below. Without unobstructed access to the levers, picking became significantly more difficult.*

ABOVE: *The bits of Hall's key are numbered for easier user reference. The bits numbered 2 and 7 are missing from this key, but with all seven in place the sides would be flush with the ends of the mounting bar. The assembled bit would be placed in the keyhole and the lock opened with the wrench.*

HALL'S CHANGEABLE BIT KEY LOCK

STERNS & MARVIN'S FIREPROOF SAFE LOCK

The firm of Sterns & Marvin was a predecessor corporation to the Marvin Co. of New York. While Sterns & Marvin primarily used well-known locks on their burglar-proof safes, they did manufacture their own locks for use on their fireproof models. One Sterns & Marvin lock was a push-key design with a removable wrench thought to date to about 1854.[61] The removable wrench meant that applying constant pressure to the bolt was more difficult, unlike other push-key designs with fixed bolt handles such as the Grasshopper. This removable wrench was to become a common feature on safe locks from both Sterns & Marvin and the Marvin Co.

ABOVE, TOP: *Large and heavily constructed for a fireproof safe lock, Sterns & Marvin's push-key lock was probably easier to pick than to force open, since the keyhole, though narrow, gives ready access to the six levers.*

ABOVE, BOTTOM: *Although the lock was sturdy, Sterns & Marvin's key was no more intricately made than contemporary push-style keys.*

OPPOSITE, TOP: *Harig & Stoy's lock required a key bit with a removable shaft and a wrench for its operation. The keyhole plate at center ensured that the levers were inaccessible from the outside whenever the bit was in contact with them.*

OPPOSITE, BOTTOM: *Harig & Stoy spared no complication in its 1854 design. There are no records as to its long-term reliability, but this lock's impressive hundred-plus-piece construction includes two coil springs for rekeying (visible at the top of the mechanism) and a large watch spring to return the traveling key bit (inside the key cylinder).*

HARIG & STOY'S BANK LOCK

Also in 1854, lock makers Harig & Stoy of Louisville, Kentucky, introduced a new changeable-bit lever lock. With more than one hundred parts, the Harig & Stoy model was a large, complicated lock for its time and required high-quality machining throughout as well as complicated operation. To open the lock, the user would insert the key and unscrew the handle, leaving the bit in the keyhole. The wrench was attached to the bit carriage and turned, moving a plate over the keyhole and the bit up to the tumblers. This same wrench is then removed and reattached to the bolt to throw it. As the bolt opens, the bit and keyhole are returned sharply to the opening by a large watch-style spring that provides constant tension against the key.[62] To rekey this lock, the lock was first opened and the wrench turned in a change-key hole in the back of the case, loosening the levers held in place by coil springs visible at the top of the case. Then with the reordered key inserted and turned partway, the wrench was removed, allowing the levers to tighten on the new key.[63] It is not known how many of these masterful locks Harig & Stoy produced, but only two are known today.

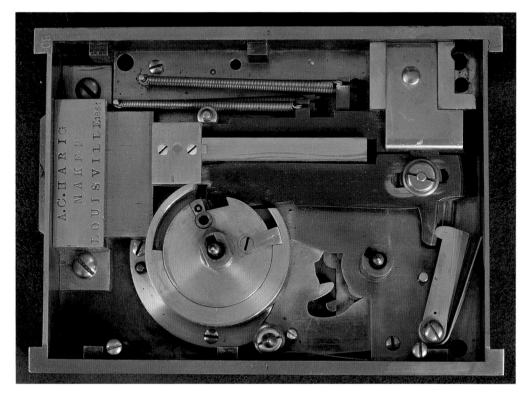

J. B. BRENNAN'S BANK LOCK

In 1854, J. B. Brennan patented a sturdy ten-lever key lock.[64] The key is a two-piece design with the rotating bit acting as a push key on the levers while allowing the flanged shaft to turn, throwing the bolt. Brennan's lock operates on a principle similar to the grasshopper locks prevalent at the time, but the union of the bolt handle and bit into a single key offered a significant improvement. The major weakness of the basic grasshopper design was the ease of access to the levers combined with the ease of putting pressure on the bolt handle. In Brennan's heavy design, the ten levers are placed behind the bolt, making it awkward for the levers to be held in position and the bolt thrown without the original two-piece key. No records of J. B. Brennan or his locks are known and this example from the Mossman Collection is thought to be the only example of his work to remain today.

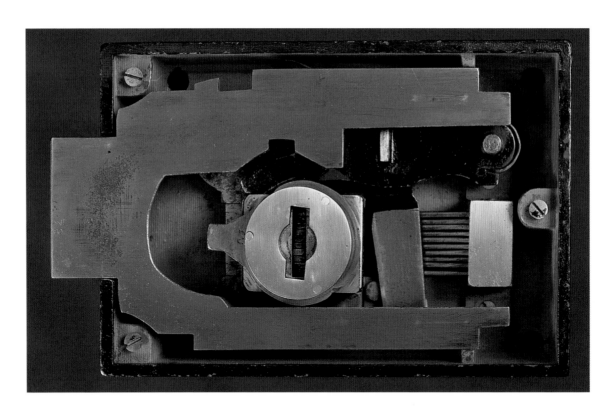

ABOVE: *J. B. Brennan's key lock. Designed for use in safes, Brennan's ten-lever lock offered quite good security for his time.*

RIGHT: *Brennan's push key was stronger than similar contemporary keys, for it acted as the wrench as well.*

LEWIS LILLIE'S CLICK LOCK

In 1855, Lewis Lillie was still early in what would be a long and significant career in the field of safes, safe locks, and time locks. His firm at that time was known as the World's Safe Company, later changed to the more decorous Lewis Lillie & Co. Lillie's first major safe lock innovation was his Click lock, a two- or three-tumbler combination lock operated by a removable key. The user sets the key in place on the spindle and pushes this spindle to the back of the case, engaging the first tumbler, and turns the key, counting the number of "clicks," hence the name. When the first tumbler is set, the key is pulled out to engage the second tumbler and turned, again counting the clicks. With only two or three tumblers, the Click lock could not afford the top levels of security against burglary even for its time, but it does represent an evolutionary step between Evans & Watson's 1852 letter combination lock and Isham's later, sophisticated, high-security Key Register lock of the 1860s. Today, two Click locks are known to survive.

Lewis Lillie's Click lock used combination lock principles for its security, but with only two or three tumblers its security was limited. The Click lock's discrete turning mechanism made turning more accurate, but with no marked dial the user was required to count the clicks.

DAVIDSON'S FIRE KING BANK LOCK

Most of the sharpest competition in the safe industry focused on the high-value area of burglarproofing, yet a greater volume of safes was sold for the ability to offer some burglarproofing and solid fireproofing. By the late 1850s many of the locks found on fireproof safes were no longer patent-protected, and though any accomplished safecracker could likely open such a lock, the contents generally did not warrant greater protection. A few safe makers focused on the fireproof safe market, including J. Davidson of Albany, New York, who would later combine with MacBride. Davidson made both fire- and burglarproof safes and also designed his own locks, including his Fire King line of safes and their specially made lock. While Davidson is not known to have gained national renown, the Albany area was quite prominent when this lock was made in the 1850s. The earliest of Davidson's Fire King safe locks featured a cast-brass plate advertising its fireproof quality, and a second, similar plate would be found on the safe exterior.

Lure of the Lock suggests that this lock is similar to Day's key lock,[65] but in fact these locks have little in common. Day's key lock (page 86) was a lever mechanism, common among fireproof safes of the 1850s. But when it was made, Day's mechanism with its Chubb-style detector was intended as a burglarproof design. The levers have antipicking teeth clearly visible on the inside of the gates and a key-and-wrench mechanism obscures the keyhole at the time the key bit contacts them. However, the length of the wrench shank is far too short to accommodate the thickness of doors that had become standard on burglarproof safes of the period. Consequently, the Day key lock is thought to have been made earlier, probably around 1840, a time when even burglarproof safe doors were much thinner. In fact, the Davidson Fire King lock has much more in common with the Solomon Andrews locks of the 1830s and '40s, including a variable number of levers and a changeable-bit key.

ABOVE: *Davidson's Fire King safe lock is easily identified by its ornate plaque. A similar, larger illustration was commonly found on the outside of the safe itself.*

OPPOSITE, TOP: *Although it was no longer the state of the art, Davidson's Fire King lock was hardly low security. With eight levers and antipicking teeth first found in Newell's Parautoptic (visible vertically, open case, center), the Fire King lock made excellent use of public technology.*

OPPOSITE, BOTTOM: *Even the Fire King's key was reminiscent of the Parautoptic's key. Despite using fewer levers, the design was popular and reliable.*

FIRE KING LOCK

DAY'S KEY LOCK

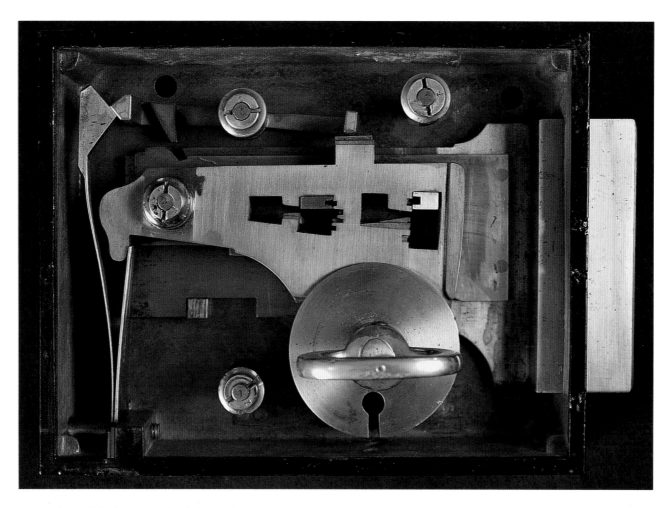

ABOVE: *Day's key lock. Day's lock bears all the hallmarks of burglarproof construction of the 1840s: ten levers, false gates (visible on the fence and lever openings), and a Chubb-style false key detector.*

RIGHT: *Day's removable-shank bit distanced the levers from the keyhole, making picking even more difficult. The shaft was unscrewed from the bit, leaving the bit inside the lock to engage the levers as the handle was turned.*

T. J. SULLIVAN'S COMBINATION LOCK

T. J. Sullivan was another designer of safe locks, assigning a number of his patents to MacBride & Davidson, which included Sullivan-designed locks on its burglarproof safes through the 1860s and '70s. Sullivan's most advanced model was a key-changing combination lock patented in 1867.[66] It featured an anti-micrometer mechanism that intermittently lifts the drop lever off the tumblers as the dial is turned. Sullivan's combination lock was a significant competitor to Sargent's Automatic, but MacBride & Davidson's safes never received the exposure that would have brought Sullivan's design the recognition it deserved.

ABOVE: *T. J. Sullivan's 1867 combination lock was a common feature on later MacBride & Davidson safes. With a change key and anti-micrometer device, it offered security on par with its more widely acclaimed contemporaries.*

RIGHT: *Change keys were a feature common to many high-quality combination locks of the time. They made altering a lock's combination quicker and cleaner than doing so by hand, and allowed owners to grant access to an open safe to employees while maintaining control over the lock's combination.*

LINUS YALE, SR.'S PEANUT KEY LOCK

By 1855 Linus Yale, Sr., was well established as a major figure in the field of lock innovation and in May he received a patent[67] for what has come to be known as his Peanut Key lock, an ingenious and unusual relative of the pin tumbler. The key, a curved brass slug with five holes of varying depth, slides lengthwise into a fitted opening at the top of the case. As the handle is turned to the right, the key is pressed against the ends of five two-piece pins, aligning their split points and allowing the locking plate to slide up, over a stump, and the bolt to slide back, into the case. As the bolt comes to rest, the recess where the key is held is brought into line with the opening in the handle, allowing the key to drop out, into the user's hand.

This design should not be confused with lower-security locks like the grasshopper lock, since Yale's Peanut Key lock presents major difficulties to the safecracker. The key recess is too small to hold enough black powder to blow the lock open and, because the key recess moves away from the insertion point before the bolt is engaged, there is no opportunity to manipulate the pins through the keyhole. But the best measure of this lock's burglarproof qualities was its use on a vault. This example was used by the Judson Bank of Ogdensburg, New York, and was given to the Mossman Collection by the National Bank of Ogdensburg.[68] At least one source identifies this as a "precursor" to the pin tumbler mechanism,[69] but the modern pin tumbler had been introduced by Linus Yale, Sr., in 1844,[70] well before the Peanut Key lock.

ABOVE: *Linus Yale, Sr.'s Peanut Key lock. The entire key is placed in the hole at the top left and is ejected through the handle as it is turned. Small and tightly built, the Peanut Key lock was surprisingly secure.*

OPPOSITE: *The back of the Peanut Key lock, shown with two keys. The holes of varying depths are clearly visible, acting as "negative" pins.*

PEANUT KEY LOCK

HENRY ISHAM'S PERMUTATION LOCK

Beginning at least as early as 1845, Henry Isham, the leading inventor at the New Britain Lock Co., was refining his Permutation lock. Over the twelve-year period ending in 1857, Isham designed at least three distinct versions, culminating in the model shown here. The earliest version of Isham's Permutation lock was based on his 1845 patent[71] that set forth the Permutation lock's major parts, the set of circular tumblers each turned by a sliding rack that was operated by a cylindrical key shaped like a winding crank. This model was likely premature for commercial introduction and is not thought to have been produced. Isham's second version, the earliest version for which an example is known, was a four-lever model that incorporated improvements from his 1856 patent,[72] including his novel rotary bit key with what Isham termed "sector pinions,"[73] layers of lateral cog teeth.[74] The third and final version of Isham's Permutation lock shares the same basic design of circular levers (visible in the upper right of the case) and Isham's rotary bit, but adds refinements from his 1857 patent,[75] including a wonderful two-piece key that combined the functions of a bolt wrench with the rotary bit. The removable rotary bit is held in place by a flange that rests in a shallow cup on the shaft end and is kept stationary on the shaft by the spring-loaded sleeve, which also supports the two pins that act as the bolt wrench.

Once the key is inserted and pushed into place, the rotary bit is released by the sleeve and is free to turn with one end supported by the shaft cup and the other by its end pin resting in a small hole visible in the plate to the lower left of the case center below the semicircular opening. As the key handle is turned counterclockwise, the wrench pins begin to move a set of intermediary tumblers, the sliding racks that run across the case bottom past the circular tumblers. These racks have tooth patterns on their top edges that engage the rotary bit, which counterrotates clockwise as the tumblers slide beneath it, but will block all movement if the sector pinions do not match the sliding racks' teeth. The sliding racks advance various distances, turning the circular tumblers to align their gates and allow the bolt to retract. Only a single example of each type of Isham's Permutation lock is known to survive.

ISHAM'S PERMUTATION LOCK

TOP: *Isham's Permutation lock was large, imposing, and nearly solid metal, with its many closely machined pieces leaving little room in the case. Despite its great mass, the mechanism moved easily due to its fine workmanship. The inside of the keyhole is visible as a crescent-shaped opening around a small hole at the left center. The stack of round tumblers can be seen to the far left, along with the toothed sliders that turned them.*

ABOVE LEFT AND RIGHT: *Isham's key used "sector pinions," a stack of partial gears with teeth extending varying distances around their circumferences. This bit remains with the shaft but rotates in the opposite direction as the shaft is turned, with the pattern on the sector pinions controlling the distance that the tumblers turn.*

KRENKEL'S KEY LOCK

From the 1860s until the 1880s, K. Krenkel was a safe and lock maker located at 198 Broome Street, New York.[76] One Krenkel design of interest is a key lock designed and possibly patented in 1856.[77] This lock has long been known by its description in *Lure of the Lock* as having been made in 1875 and intended for a fireproof safe. By 1875, with high-quality combination locks available, a design such as Krenkel's would, in fact, qualify as little more than an interesting fireproof safe lock. However, with thirteen levers, Krenkel's key lock was a daunting burglarproof lock when it was actually introduced in the 1850s. Further, the key was not a simple flat key, as described in *Lure of the Lock*.[78] Rather, Krenkel's key has a subtle curvature that runs the length of the key, and it was one of the first locks to make use of both sides of its key.

Like Yale's third type of Magic Key Lock, Krenkel's key lock also used two sets of levers acting on either side of a key. However, with its simpler construction, Krenkel's lock was far less expensive. The thirteen triangular levers are held in a single alternating stack around the handle, with seven acting on top of the key and six on the bottom.

OPPOSITE: *Krenkel's key was not symmetrical and included a slight curvature to ensure that it was inserted correctly. Since the levers were held in a single alternating stack, the bits are offset on the top (left) and bottom (right) sides.*

CHAPTER 3

The Combination Lock Comes of Age: 1857–1871

The middle of the nineteenth century saw a major change in bank security throughout the United States: a shift away from key locks to combination locks for the most important applications. The idea for a combination-based lock was not new, but machining tolerances had never allowed for a design that could offer safety on par with the best key locks. With improved manufacturing techniques, the combinaton lock quickly became the equal of the key lock and the ease of maintenance and changing combinations made it popular with bankers. In Europe, however, bankers and manufacturers would not adopt the combination with the same zeal, continuing to rely on ever more complex key locks well into the twentieth century. But for bankers in the mid-nineteenth-century United States, the combination lock was quickly becoming the standard of security.

HENRY COVERT'S COMBINATION LOCK

In the history of bank locks, Henry Covert usually commands little attention. He was, however, the coinventor (with the famous James Sargent) of both the Magnetic Lock of 1865 and the key-changing feature for combination locks in 1866. Covert was also the designer of a quite successful combination lock design based on three of his own patents granted in 1857, 1858, and 1859.[1] A hand-changing design, Covert's patent combination lock featured a pull-out dial anti-micrometer mechanism that guaranteed that the dial was never in contact with both the tumblers and the fence at the same time. After dialing the proper combination, this dial is pulled out, raising the fence into the tumblers and allowing the bolt to be thrown. Though the example seen here has a dial numbered through 100, Covert's patent was one of the few combination locks commonly produced with a dial numbered through 200, an option that required greater than average accuracy, both on the part of the maker and on the part of the user.

Covert's patent was constructed and sold by a number of makers with various numbers of tumblers and in various sizes under license from Covert. *Lure of the Lock* identifies this example as one made by Briggs-Covert,[2] and although Martin Briggs of Rochester, New York, was one of two major producers of Covert's patent, the example shown here was made by the other, W. W. Bacon of New Haven, Connecticut. Approximately twenty-five examples of Covert's patent combination lock are thought to survive today from various makers.

Covert's patent combination lock was among the most finely machined locks of its day. The hand-changing anti-micrometer design was available with a dial numbered through 200, making it more difficult to open than other locks for both thieves and owners.

AMOS HOLBROOK'S TIME LOCK

The year 1858 saw the debut of the first time lock to be used in a bank, the Holbrook Automatic Lock. Based on an 1857 patent,[3] this time lock was first mentioned in a *Milford Journal* article[4] that year, and by January of 1858 an improved version of the Holbrook Automatic[5] was in use in at least two locations, the Milford Bank of Milford[6] and the Holliston Bank of Holliston,[7] both in Massachusetts.

The oft-repeated and widely held belief that the first time lock ever purchased and used by a bank was made and installed by Sargent & Greenleaf in 1874 seems to stem from a 1927 letter from the First National Bank of Morrison, Illinois, to Sargent & Greenleaf, Inc. A portion of this letter was reproduced in Sargent & Greenleaf's 1927 sales catalogue[8] and in the *Lure of the Lock* in 1928,[9] eventually giving rise to a general acceptance of Sargent's priority.[10] But the 1927 letter was written more than fifty years after the fact at a time when the popular understanding of both the history of the earliest time lock makers and the details of the litigation surrounding this industry was already clouded. For Sargent & Greenleaf's advertising purposes, it was no doubt convenient to forget that, regardless of Holbrook's ultimate financial success or failure, the 1858 adoption of the Holbrook Automatic—sixteen years prior to the earliest Sargent & Greenleaf design—firmly established this lock as the first commercially used time lock.

Holbrook used two separately removable seventy-two-hour movements,[11] each with its own twelve-hour enamel dial on a plain brass front plate. The case is heavy cast iron with black and scarlet paint. The mechanism is covered by a sliding panel, shown open, moved with a small brass knob. The lock uses two bolts, retracted here, which when set extend out of the case to the right. Four coil springs, visible at the right interior of the case, keep inward tension on the bolts and are allowed to pull the bolts into the case when the timer runs down. This design proved quite reliable; the Holbrook Automatic in the Milford Bank operated for at least eight years with no reported problems.[12] Holbrook would eventually sell one half of his interest in his Automatic lock to Samuel W. Hayward around 1860[13] and assign the other half to his wife, who had financed the time lock's development. Holbrook's wife sold this interest to Hayward for $1,000. Eventually, Hayward would sell all the rights to the Holbrook Automatic to Yale through Thomas Keating in 1876.[14]

Court testimony by Amos Holbrook himself suggests that between twelve and eighteen Holbrook Automatics were made,[15] although the number may well have been as great as thirty.[16] While Holbrook's own estimate would normally be persuasive, at least one other Holbrook statement—that only two of his Automatics were installed[17]—is contradicted by both a deposition of Joseph Hall[18] and a *Milford Journal* article.[19] In an 1879 deposition, Joseph Hall identified a Holbrook Automatic installed by his company for First National Bank of Milford in June 1875, less than one month before Hall's Safe & Lock would offer a time lock of its own. And in what may be the first media report on the effectiveness of a time lock, the

Each of the two clock movements in Holbrook's Automatic could be individually removed for repair or maintenance. This semimodular movement feature would not be seen on major time lock models until 1878. The coil springs at the right of the case acted as a light bolt motor, retracting two bolts into the case.

OPPOSITE: *The original label found on the sliding door on Holbrook's Automatic. The term "automatic" did not yet refer to a bolt motor specifically and was used here in its more conventional meaning: the lock opened without any action on the part of the user.*

Holbrook's Automatic Lock was the first commercially used time lock. Compared to later time locks its appearance was spartan, with its black cast-iron case, sliding door, and unadorned mechanisms. However, Holbrook's Automatic had all the necessary elements that would become standard on time locks.

Milford Journal reported that a Holbrook Automatic foiled a burglary attempt at a bank in Great Barrington, Massachusetts. Further, Samuel Hayward stated in a deposition that thirty time locks were made, that he sold eighteen to twenty of them, and that he knew of five that were installed in banks.[20]

Though we find contradictory evidence about the specific success of the Holbrook Automatic, it is clear that the lock gained only limited exposure and ultimately was not a financial success.

Only two examples of the Holbrook Automatic Lock survive today. Plain and utilitarian by comparison to later time locks, the Holbrook Automatic Lock is now recognized as the first commercial time lock.

JOHN P. LORD'S GEAR LOCK

In 1858, John P. Lord of Manchester, New Hampshire, was awarded patent no. 21,346 for a new combination lock.[21] Dubbed the Gear Lock by Lord, it is operated by a handle-like key that turns five concentric ring-shaped tumblers through a gear that engages the outermost ring. When the tumblers are aligned, the key can then engage and throw the bolt.

With further refinement and sufficient financial backing, this lock could probably have been a significant success, considering the later popularity of the Dodds "Eureka," or Treasury Lock, introduced in 1862 with a similar basic design. This is thought to be the only surviving example of Lord's Gear Lock.

Lord's Gear Lock introduced a simple but effective concentric tumbler design. The combination could be changed by adjusting the location of the square-headed screws visible in between the tumblers.

L.H. MILLER'S BANK LOCK

Also in 1858, a bank lock was patented by L. H. Miller[22] of Providence, Rhode Island, featuring nine levers designed to be inaccessible through a minuscule keyhole. A thin tube-shaped key is visible in the center of the bolt handle, which when in place allows the interior mechanism to avoid the tumblers as it is turned. The levers are unmeasurable, even with pressure applied to the bolt, and can be easily loosened and rearranged when the lock is open, as seen below.

ABOVE: *The lever-case cover of Miller's lock.*

LEFT: *Miller's lock with its lever-case cover off.*

OPPOSITE: *L. H. Miller's lock with its tiny, tubular key visible in place at the center of the actuating handle. The minuscule keyhole was designed to offer no room for a picker to detect the levers.*

LYMAN DERBY'S COMBINATION LOCK

The safe making firm of Marvin & Co. would eventually become part of Herring-Hall-Marvin, but it was still independent in 1860 when employee Lyman Derby was awarded a patent[23] for his combination lock. A lock of equally complicated construction and operation, Derby's patent uses a single knob to operate the four tumblers or the bolt depending on the set screw visible at its center. With the knob indicator set to "A," this screw is loosened to free the knob from the bolt. The knob is then turned counterclockwise one full turn, continuing to the first combination letter, and then returned to "A," allowing an internal pin to drop past the first tumbler to the second. The knob is again turned counterclockwise once, continuing to the second letter, and then returned to "A," setting the second tumbler. The third and fourth tumblers are set similarly, except that the initial counterclockwise motion is two and three turns, respectively. Finally, the set screw locks the knob, which is turned three times clockwise to throw the bolt.[24] The only known example was possibly a salesman's sample, as it would have been unusual for a production lock to prominently identify both the maker and the patenter, as seen on its dial collar. The example of Derby's lock was donated to the Mossman Collection by Diebold Lock & Safe Co.[25]

ABOVE: *The letter dial of Derby's patent combination lock with the indicator set between K and L. The set screw in the center of the knob engages and disengages the bolt.*

OPPOSITE: *The complex asymmetric construction of Derby's patent lock is evident in the left and right views of the interior of the only known example of this lock.*

LINUS YALE, JR.'S DOUBLE TREASURY LOCK

By 1860 obsolescent key-based designs were already giving way to combination locks, and only the most intricate key locks were still considered for roles requiring top security. Consequently, when Linus Yale, Jr., presented a new key lock for use by the United States Treasury Department, it was unquestionably among the best ever in both design and execution. Adopted by the Treasury Department, the lock would become known as the Double Treasury and cement Yale's name among the premier bank lock designers of all time.

Yale's Double Treasury was first patented in 1860[26] and quickly improved in 1861.[27] It incorporated a number of security measures found in earlier Yale designs, the most important of which was the key bit. The seven interchangeable levers were operated by a changeable bit that detached from the shaft, and as the shaft was turned the bit was carried away from the keyhole to the tumblers. This made the tumblers all but inaccessible through the keyhole and allowed the Double Treasury to use the key shaft itself to actuate the movement, eliminating the need for a second handle or wrench. New to the Double Treasury was a "disconcerter," a device that prevented even a specially engineered false key from taking an accurate impression of the tumblers.

This lock seems to have been available in both single and double formats, but it was the single-custody two-mechanism version shown here that would be adopted by the government and gain renown. Although Yale's Double Treasury lock was quite successful, the actual extent of its production is not clear. Today, fewer than five examples are known.

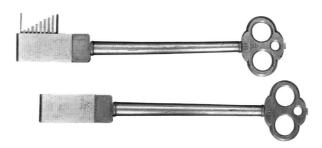

ABOVE: *Yale's Double Treasury key, shown here with its bit fully hidden as it would be carried (below) and partially extended out of its slot (above). As the key was turned, the bit exited its slot, traveling into the lock where it engaged the levers.*

OPPOSITE: *By the time Linus Yale, Jr.'s Double Treasury key lock was adopted by the U.S. Treasury Department, key locks were already obsolescent. Within five years, bank lock development in the United States would be focused primarily on combination designs. However, Yale's design, with its changeable, detaching bit and the addition of a "disconcerter" to prevent the accurate detection of tumbler positions, made this lock one of the most secure key locks ever.*

MARVIN SAFE CO. SALESMAN'S SAMPLE

Just as bank lock and time lock makers made especially ornate samples of their premier pieces for exhibitions and their sales staff, some safe makers went to the same extent to introduce their products to a target market. Safe makers would sometimes construct entire safes in miniature, and one such model safe is known from the Marvin Safe Co., but more common was the sample vault door.

This salesman's sample vault door was made by the Marvin Safe Co. around 1860 and is an excellent guide to the construction that made these vaults so secure. By the time this design was in production, the use of explosives such as black powder or nitroglycerin to blow off vault doors had become a well-known burglar's tactic. In this model, we see the layered construction of the door edges that guaranteed such a tight fit that the door would have to be closed with a compression bar.

Such a tight-fitting door meant that a burglar could not introduce a sufficient volume of explosive into the door-safe gap to rupture the hinges or bolts. The compression bar is included on what would be the outside of the door. The boltwork can be examined closely and clearly shows the accommodations for two independent combination locks, for single or double custody.

Salesmen's samples are quite rare in general, and sample safes and vault doors even more so. Surviving examples are often of the more widely produced low-quality samples that showed the basic workings but did not incorporate the same materials or workmanship. This sample Marvin vault door is the only known model of such high quality and detail throughout.

A salesman's sample of a Marvin Safe Co. vault door. The model was made to the same exacting tolerances as the actual doors, including the six-tier doorjamb and functioning compression bar. Though the model hinges are attached with screws rather than with the bolts that would be used on an actual vault, Marvin's boltwork secures the door on all four sides, so detaching the hinges would be of little use.

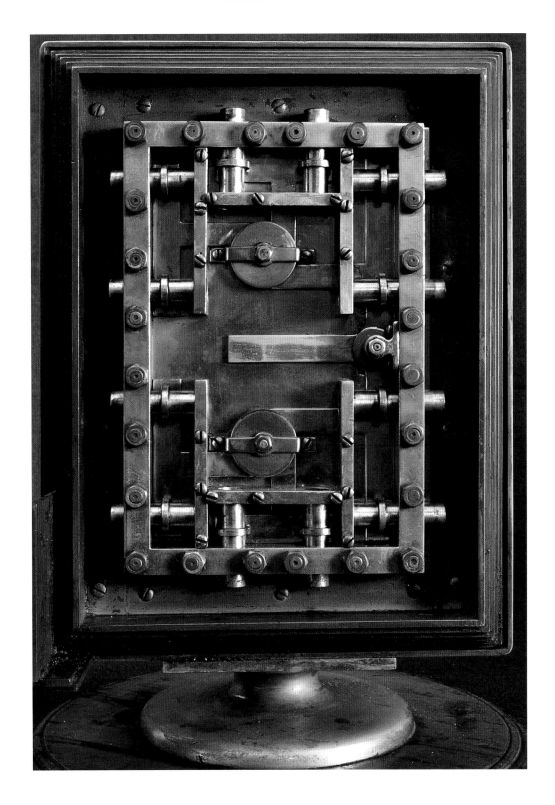

MARVIN SAFE CO.

Engraving of Marvin Safe Co.'s factory. This Marvin catalogue illustration from 1889 outlines the production process for fireproof safes at the company's Thompson Street, New York, location.

W. B. DODDS TREASURY LOCK

In 1862, W. B. Dodds Co. debuted a new combination lock that would continue to be produced for the life of the company and for a brief period by Dodds-MacNeale-Urban. The dial on this five-tumbler alphanumeric model is marked with numbers 1 through 9 interspersed among twenty-three letters, A through Y excluding I and U. The hand-changing tumblers are concentric cylinders, each with a pair of small protrusions allowing each of the four interior tumblers to turn the one outside it, much like stacked combination tumblers. When aligned, gaps in the tumblers allowed the bolt to slide through. Originally dubbed the "Eureka" lock, Dodds advertised this design as superior thanks to its reliability from a springless, gravity-independent design.[28] This design eventually became well known as a model adopted by the U.S. Treasury Department[29] and is today commonly referred to as the Dodds Treasury lock. Fewer than five examples of this lock remain today.

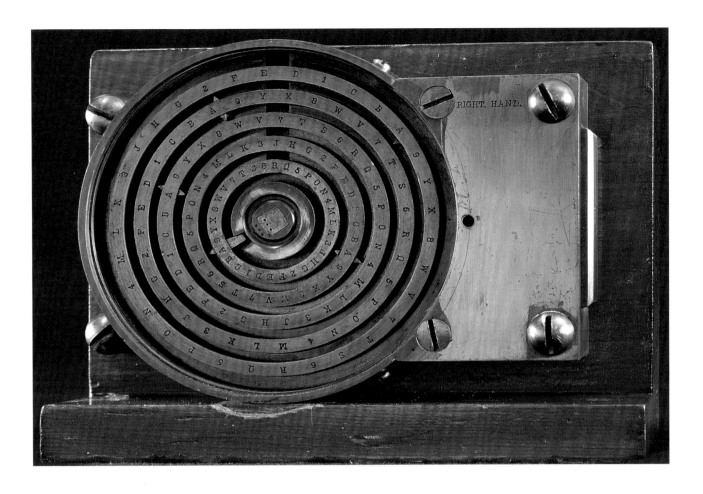

Shown here with its cover on and off, the Dodds Treasury lock was originally named the "Eureka" lock. Its mechanism does not rely on springs or gravity, making it highly reliable, while its more than thirty million combinations and tight construction made it very secure. The five concentric ring tumblers are held in place by the lid, making changing the combination simple in theory but awkward in practice.

LINUS YALE, JR.'S DOUBLE DIAL BANK LOCK

With the introduction of his Double Dial bank lock in 1863, Linus Yale, Jr., altered the direction of the bank lock industry in the United States and set a new standard for vault security. The Double Dial had two sets of tumblers known as curbs or wheel packs that were locked in place by a pair of five-pin feather key pin tumbler locks, also a recent Yale invention.[30] Each three-tumbler combination lock was capable of one hundred million combinations and could have the combination changed in two ways, either the regular realignment of the tumblers by hand or by adjusting where the dial engaged the wheel pack. This second, quick method changed the combination numbers, but retained the numbers' relation (i.e., 10–20–30 could easily become 11–21–31). The Double Dial could be set for either single- or double-custody operation and was available with offset spindles. It also featured a disconcerter, an anti-micrometer device that keeps the drop lever from contacting the tumblers until all three have been correctly set.

Yale's Double Dial was awarded a silver medal at the Paris Exposition of 1867 and the most expensive version of the lock seen here carried a silver-plated replica of the obverse and reverse of the Paris Exposition medal on the wheel packs. According to Yale's 1872 price list, the version of his Double Dial shown here sold for $350. By 1883 competition had driven the price down to $180, but Yale's Double Dial would remain a significant model for Yale until the 1890s, ending by 1896.

LEFT AND OPPOSITE: *Linus Yale, Jr., was best known for his key locks, but the combination lock revolution of the mid-1800s did not pass him by. Yale's Double Dial design won a silver medal at the 1867 Paris Exposition. Yale included the two sides of that medal on the wheel pack covers of the most expensive version of his Double Dial.*

ABOVE: *Linus Yale, Jr.'s feather key. A pair of locks using these five-pin keys secure the Double Dial's wheel packs. Yale's feather lock and key is the forerunner of all modern pin tumbler locks. Its thin, convenient key was a true innovation at that time.*

MacNeale & Urban's Excelsior Combination Lock

Safe maker MacNeale & Urban was a major supplier of the combination lock design of W. B. Dodds but also designed and sold its own locks. In 1864, MacNeale patented the Excelsior,[31] a simple combination lock with an interesting key mechanism for changing the combination. This design was revamped around 1870, making the Excelsior smaller and less expensive to produce. Both the early and later Excelsior models used an indirect-drive mechanism with the dial operating a set of gears that in turn aligned the tumblers. But unlike later indirect-drive locks that placed the spindle outside the case to prevent a thief from gaining access to the lock's interior by pulling out the dial and spindle, the Excelsior used this interior version of indirect drive to allow for a quick and simple method to change the combination. This relied on the thumbscrew on the top of the early model and the large set screw in the lower center of the later one. These screws, when turned by a change key, allowed the tumbler stack to slide slightly when unlocked, disengaging them from the spindle gears and allowing a new combination set by clearing the sixty-digit dial and then entering the new combination.[32] The Excelsior was available with up to six tumblers and also in dual-custody double-dial models.[33] Today, fewer than five examples of MacNeale's Excelsior combination lock remain.

The Excelsior combination lock from MacNeale & Urban used a geared or indirect-drive design as part of its combination-changing method rather than as an independent security feature, as in later locks. When inserted, the Excelsior's change key allows the user to clear the old combination by spinning the dial and then simply dialing the new combination.

The Excelsior's change key.

116

TILTON, McFARLAND & CO.'S BANK LOCK

Tilton, McFarland & Co. was a safe making firm that offered both fireproof and burglarproof models. Though its burglarproof safes generally featured locks by other makers, Tilton, McFarland developed and used its own safe locks in the fireproof models. Unlike other makers, Tilton, McFarland did not patent these fireproof safe locks, instead probably relying on their low cost, modest security, and inaccessibility inside the safe door to avoid duplication. An example of a Tilton, McFarland fireproof safe lock from about 1865 uses both a combination lock and a key lock to offer sufficient fireproof security. The combination must first be set, then the key can unlock the safe door.

Tilton, McFarland & Co.'s fireproof safe lock used both a combination and a key lock and it was sealed entirely into the safe door rather than mounted on the inner side. Because fireproof safes such as these were sold primarily to individuals and small businesses, there was little chance that the safe door would be broken open and the lock copied.

JAMES SARGENT'S MAGNETIC COMBINATION LOCK

Throughout the nineteenth century, European financial institutions had equally well developed safe and vault technology as that of the United States. However, while European inventors continued to introduce innovative and elegant key locks, key lock development had substantially ended in the United States by the 1860s. The broad acceptance of the dial combination mechanism throughout the American safe and vault industry offered new security challenges to safecrackers. With no keyhole, safecrackers increasingly turned to drills, wedges, torches, and the like, but those intent on defeating the combination lock found that early designs were vulnerable to new methods of attack. The image of the stethoscope-wearing safecracker is likely fictional, since even the earliest combination locks were not so loosely made that the fence could be heard on the tumbler gates. Rather, safecrackers found that pressure applied to the bolt handle, forcing the bolt against the tumblers, allowed the manipulator to determine the tumblers' positions as the dial was turned. James Sargent refined this cracking technique in the mid-1860s into the micrometer using a weighted armature that attached to the bolt handle, magnifying even minuscule movements and allowing all but the most exactly machined locks to be opened.

Using this objective test of the weakness of the then-state-of-the-art combination locks, James Sargent developed and patented his Magnetic Lock in 1865,[34] which instantly placed him among the foremost bank lock makers. Sargent's Magnetic Lock featured a powerful bar magnet visible above the fence and the fence bar. This magnet held the fence up, off of the tumblers except for a small portion of the dial's rotation, making it nearly impossible for even a micrometer to distinguish the tumblers' positions. The Sargent Magnetic also introduced a new change key mechanism that required a second combination be dialed before the change key could be used to alter the combination. This combination-released change key mechanism would become an almost universal feature of high-quality combination locks for the next century.

While the power of the magnet would wane over time, eventually allowing the fence to rest on the tumblers, remagnetizing this piece was a relatively simple matter. Of this earliest version of Sargent's Magnetic Lock featuring a sliding bolt, only one example, removed from the Bowery Savings Bank in New York,[35] is known to survive.

A schematic of James Sargent's micrometer. The armature (center) was attached to the safe's bolt handle (right) with sufficient weight to keep the bolt forced against the tumblers. The length of the armature magnifies the subtle motions of the bolt as the combination dial turns the tumblers and registers these on the dial. Far more sensitive than even the most trained fingers, Sargent's micrometer allowed him to defeat many of the most secure combination locks of the day and opened the door for his Magnetic and Automatic locks.

Change key for Sargent's Magnetic Lock. Sargent was awarded a patent for his combination change key, which revolutionized how combinations were changed.

James Sargent's Magnetic was the lock that made him a major success. The finely machined and reliable key-changed combination design was improved by the addition of a powerful bar magnet above the tumblers that kept the fence from touching the tumblers for all but a moment, rendering even the micrometer ineffective. Although the magnet's power would wane over time, the balance of Sargent's lock was still a formidable device.

JAMES SARGENT'S ROLLERBOLT MAGNETIC COMBINATION LOCK

Throughout 1865 Sargent advertised both his ability to open his competitors' combination locks with his micrometer and his Magnetic Lock's anti-micrometer design. In 1866, Sargent introduced an improved Magnetic with a rotating bolt or "rollerbolt" that moved the dogging action of the lock from the tumblers themselves to the fixed bolt axle. Sargent's combination of the magnetic mechanism, the combination-released change key, and the rollerbolt made his second version one of the highest-security combination locks ever made. Sargent sold this lock in iron for $250 and in bronze, as shown in this example taken from the vault of Chemical Bank in New York, for $300.[36] Five examples of this later rollerbolt version of James Sargent's Magnetic Lock are known today.

OPPOSITE, TOP: *The patent information for Sargent's improved Magnetic is shown prominently on the lock's most important innovation, the rollerbolt, itself. The rollerbolt would be a major design element in Sargent's combination and, later, time locks for many years.*

OPPOSITE, BOTTOM AND ABOVE: *James Sargent's improved Magnetic combination lock introduced the rotating "rollerbolt" that turned when unlocked to align a square cutout section with the side of the case, freeing a portion of the safe's boltwork to move into the case. The rollerbolt's axle isolates the wheel pack from the lateral pressure of the door bolt, ensuring that the bolt handle cannot be used to pressure the tumblers and give away their position.*

JAMES SARGENT'S AUTOMATIC COMBINATION LOCK

In 1868, James Sargent introduced his Automatic Lock, a successor to his Magnetic Lock and a major improvement on what was already one of the most secure designs known. The Automatic did away with the magnet, substituting an assembly of three levers in its place. On the back of the tumbler wheel pack is a drive cam bushing that, as the dial is turned, lifts the first lever as the bushing passes under a curved spur that extends down in front of the tumblers. When raised, this first lever raises the second lever behind it, and at the apex of the first lever's travel, when the round bushing is under the lowest part of the spur, the second lever's hook (visible at its left, under the set screw) releases the steel pin connected to the fence, allowing the fence to touch the tumblers. Should the combination be set, the fence will drop in and the bolt will be thrown. Without the tumblers aligned, the fence will touch the tumblers' edges only briefly but be lifted away again by the returning hook as the drive cam bushing is turned out from under the main lever spur. Combining the rollerbolt and the combination-released change key from the Magnetic with the Automatic device, Sargent's Automatic Lock was both a masterpiece of machining and the picture of simplicity for the user, and it quickly became the standard by which bank locks were measured.

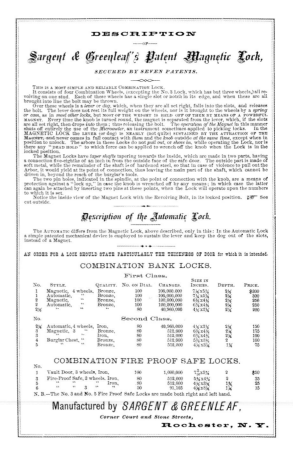

Sargent's Automatic Lock introduced a pair of levers actuated by a bushing attached to the wheel pack. The primary lever is visible above the bolt and wheel pack with a spur extending down toward the bushing fixed on the wheel pack by a set screw. The second lever that holds the fence is visible, in part, behind the primary one, with its hook securing a pin connected to the fence. This mechanical solution to suspending the fence did away with the magnet and reduced maintenance requirements. (Courtesy of Mark Bates Collection.)

OPPOSITE: These instruction sheets for Sargent's Magnetic and Automatic locks were provided to dealers and installers, describing the mechanisms, their use, and listing wholesale prices for the numerous styles.

NEW BRITAIN BANK LOCK CO.'S ISHAM-PILLARD BANK LOCK

In the late 1860s, the New Britain Bank Lock Co. was still a major bank lock manufacturer and in 1867 the company introduced a double bank lock that combined two of its most popular mechanisms from its two best-known engineers. The lock united Oliver Pillard's dial combination lock with Henry Isham's Key Register lock in a single case, featuring brilliant construction throughout and a specially engraved back cover plate.

Isham's Key Register lock (the lower mechanism) was not new at this time, having been introduced at least as early as 1858.[37] According to one account, Henry Isham overheard the offhand remark by the president of the Vermont Bank around 1848 that "no man can construct a lock that another cannot pick."[38] By 1856 Isham had the basic design of the Key Register lock, which did away with a key and keyhole, replacing them with a wrench and a "key register,"[39] a toroid case containing adjustable lettered rings that correspond to the combination-style tumblers inside the case. The stack of rings creates a series of grooves through which a flange on the wrench navigates, turning left to a stop, then moving in to engage the second tumbler, then turning right to a stop, then in, and so on, until the last turn retracts the bolt. To lock, the wrench was worked out in reverse, an action that also set the combination inside the key register, making it no more difficult to change the combination than it was to work the lock. But Isham's tightly machined tolerances combined with the small, round wrench handle seems to render Key Register locks difficult to operate if they are not regularly maintained.

In January of 1857, Frederic North of the New Britain Bank Lock Company purchased all Isham's rights to the Key Register and brought him on as an employee. Before production of the lock had gone far, however, North and Isham found that Stewart Perry of Newport, New York, had patented[40] his own keyholeless lock. Although different in design, Perry's patent used a sufficiently similar key to pose a threat. With no known manufacturing interest of his own, Perry sold his patent rights to North, who then began major

ABOVE AND OPPOSITE: *The New Britain Bank Lock Co. introduced this bank lock that combined two popular mechanisms into one formidable device. Both Oliver Pillard's combination lock on top and Henry Isham's Key Register lock were individually successful, and each was sufficiently secure to be used alone. Together, they provided a nearly impenetrable barrier.*

LEFT: *The internal cover for this New Britain bank lock features intricately detailed engraving uncommon even among the flagship lock designs from this era.*

RIGHT: *The key for Isham's Key Register lock was a small doughnut-shaped pack of five discs (top left) that controlled the user's ability to turn the attached handle (top right). The handle would be turned in one direction until it stopped, pulled out, then turned back to a stop. This was repeated until each tumbler had been set. The discs were lettered (bottom left) for ease of setting the tumblers, but these letters were invisible during operation, covered by a small plate held in place by a notched disc on the outside of the case (bottom right).*

production of the Key Register lock in 1858.[41] The earliest of these carry the stamp "Perry's Patent" with his patent date, and were possibly made during the period of North and Perry's negotiations. Most commonly made as a stand-alone lock, approximately thirty examples of Isham's Key Register lock remain. Some dual-custody Key Register and Pillard combination models were made in which the two locks each have an independent bolt; in the example shown here, one lock controls the other.

LEWIS LILLIE'S PULL-OUT DIAL COMBINATION LOCK

At least as early as 1868, Lewis Lillie had moved his bank lock production facilities to Troy, New York. James Sargent's introduction of the micrometer essentially required all combination lock makers to develop a mechanism to prevent the micrometer from measuring an interaction between the bolt and the tumblers. Lillie's anti-micrometer design retained many features common to his other models: iron case, bronze mechanism, nickel-plated brass dial, and the moderate security of three tumblers. However, Lillie's addition of a pull-out dial mechanism required the user to align the tumblers with the dial depressed and then engage the bolt after pulling the dial out, disengaging the tumblers. With the dial engaging either the tumblers or the bolt, a safecracker cannot turn the tumblers with Sargent's micrometer pressuring the fence onto the tumblers; in fact, until the tumblers are set and disengaged, the bolt handle will spin freely. Due to its unusual upright fence, visible along the left interior of the case, the lock employed a leaf spring to actuate the fence rather than gravity. This constant spring pressure on the fence would have rendered Lillie's 1868 combination lock unusually susceptible to the micrometer were it not for the pull-out dial.

Lewis Lillie made key, combination, and time locks during his career. Lillie's combination locks commonly featured soundly made mechanisms but only three tumblers, as in this one. One unusual design aspect was the vertical fence, visible along the left of the case, rather than the common horizontal, gravity-actuated fence in other combination locks. Lillie's vertical fence required a leaf spring to drive it into the tumblers.

The dial on Lillie's combination lock needed to address the weakness created by the leaf-sprung fence. With a spring holding the fence against the tumblers, a micrometer could measure the tumbler positions if the bolt could be forced back against the fence. Lillie's dial featured a pull-out mechanism where the user would dial the combination, then pull the dial out, disengaging the tumblers and engaging the bolt. A dial that could engage either the bolt or the tumblers, but not both, alleviated the risk of manipulation.

OLIVER PILLARD'S COMBINATION LOCK

Beginning in 1868, the New Britain Bank Lock Co. offered a new combination lock mechanism designed by one of its leading inventors, Oliver Pillard. The Pillard combination lock was available as either a single-dial model or a double-dial single-custody model. The cases are cast bronze and the construction is of the highest quality throughout. Despite James Sargent's micrometer being well known by this time, Pillard chose a design without a specific anti-micrometer device. Consequently, these four-tumbler mechanisms were machined to the closest tolerances possible. While the double-dial model is hand-changed, the single-dial model incorporates James Sargent's combination-released change key, a feature that may have been the subject of patent infringement claims by Sargent, limiting production. Today, five examples of the single-dial model and two examples of the double-dial model are known to survive.

PILLARD'S COMBINATION LOCK

Oliver Pillard's 1868 combination lock design was available from the New Britain Bank Lock Co. in both single- and double-dial formats. Like many locks from the New Britain Bank Lock Co., it featured intricate and beautiful engraving both on the mechanism and on the cover. Although James Sargent's micrometer was well known by this time, Pillard did not include a specific anti-micrometer mechanism. Instead, Pillard and New Britain relied on extremely exact machining to render the micrometer useless. Consequently, Pillard's 1868 combination locks were among the finest ever made.

HALL'S SAFE & LOCK CO.'S CRESCENT COMBINATION LOCK

In 1869, Hall's Safe & Lock Co. produced its first Crescent double lock based on a patent by Henry Gross,[42] one of their foremost inventors. With two independent seventy-number five-tumbler combination locks controlling the bolt, the Crescent could be set for either single- or double-custody operation.[43] To open when double custody is chosen, the first combination is set and left on the last number. After the second combination is set, both dials are turned simultaneously to throw the bolt.[44] Common production models featured a bronze case and screwed-on back plate with a nickel inset medallion of Hall's profile.[45] While hundreds of these were made, fewer than twenty are known to survive. The ornate manufacture of the example shown here from the Mossman Collection suggests that it was intended as either an exhibition model or a salesman's sample. It features inlayed marquetry on the wooden mounting block, a silver-plated case, and a hinged glass back displaying garnet studs on the wheel curbs. Like Hall's earliest time lock, this example of Hall's Crescent combination lock may have been made prior to the patent date that eventually appeared on the bolt.

CRESCENT COMBINATION LOCK

Like the combination lock design of Oliver Pillard introduced the year before, the Crescent combination lock introduced by Hall's Safe & Lock in 1869 relied on its high manufacture quality rather than on any particular mechanism for defense against Sargent's micrometer. Available in double-dial format only, Hall's Crescent featured five seventy-number tumblers and very close tolerances, placing it among the most secure locks of its time. This exhibition model includes a number of aesthetic ornaments that would not be found on a production model.

HALL'S SAFE & LOCK CO.'S PREMIER SAFE LOCK

Hall's Safe & Lock Co. began producing its Premier safe lock in 1869. The Premier was one of the finest safe locks ever made and remained in production for more than thirty years, first by Hall's Safe & Lock and later by Herring-Hall-Marvin. The earliest Premiers did not have the silver portrait medallions on the tumbler wheel packs that would become the Premier's trademark feature, and that portrait medallion would again disappear under Herring-Hall-Marvin. The double-dial model of the Premier carried the hefty wholesale price of $500.[46]

On the model shown here, a lever lock prevents unauthorized access to the wheel packs. The time lock, a Consolidated Infallible not available until 1882, controls both combination locks, but only the left combination lock can actuate the Infallible's secret combination. All versions of the Premier combination lock could be fitted or retrofitted with a time lock, making it one of Hall's most successful designs with thousands sold in four different sizes. Of these, a few hundred of all the different sizes survive today. The largest format and the double-dial format are quite rare, with fewer than ten of each known.

PREMIER SAFE LOCK

Hall's Premier combination lock. Along with its Crescent combination lock, the Hall's Safe & Lock Co. also introduced what would become the company's flagship bank lock in 1869, the Hall Premier. Available in four sizes and in single- and double-dial formats, the Premier was one of the most successful designs ever, with thousands of examples made over more than thirty years. Despite this long production period, only a few hundred examples survive today. Most Premiers featured the silver medallions on the wheel pack depicting Joseph Hall. The double-dial Premier seen here is fitted with a Hall Infallible time lock.

ABOVE: *Interior view of Hall's Premier. With the back cover and wheel packs removed, the Premier's bolt and fence mechanism can be clearly seen. Also visible is the additional "Infallible" emergency fence (nickel-plated lambda-shaped piece) that connects the combination lock to the time lock. This allowed a failed time lock to be released with a secret combination.*

HERRING & CO.'S DEXTER COMBINATION LOCK

In 1869, Herring & Co. began production of a combination lock designed by Dexter in both single- and double-dial formats. Featuring a key-changing four-tumbler mechanism, the single-dial version became Herring's standard bank lock and was always installed in pairs, with boltwork connections available for use in single or double custody. The steel rod that protrudes below the bolt aids in mounting alignment. Commonly cited as one of the most aesthetically pleasing lock designs, the Dexter is primarily of bronze construction with grapevine engraving on major surfaces. Over the twenty-plus years that this model was produced by Herring, thousands are thought to have been made, but today only about fifty are thought to survive.

DEXTER COMBINATION LOCK

Herring & Co.'s Dexter combination lock. Patented in 1869, the Dexter is commonly considered one of the most beautiful and elegant mechanisms ever produced. Always installed in pairs, the Dexter included an anti-micrometer mechanism, a two-piece lever visible at the bottom left of the case. The bolt (arced over the tumblers) is held away from the wheel pack by the bushing on the lever's top left until the drive wheel's comma-shaped cutout reaches the bushing at the lever's bottom right. At all other times, pressure on the bolt can sense only the drive wheel, never the tumblers.

DEXTER COMBINATION LOCK

A double-dial version of the Dexter combination lock was available for safe and vault doors that had a sufficiently spacious boltwork to allow for a larger lock. Though operation remained the same, the mechanism was somewhat modified. The bolt's elegant curve is gone and the anti-micrometer lever has been replaced by an adjustable friction plate with a steel insert, visible resting on top of each of the drive wheels. The actual fence that drops into the tumblers is obscured from view, behind the friction plate. Based on the same 1869 Dexter patent, this example was made later by Herring, Farrel & Sherman.

JACOB WEIMAR'S COMBINATION LOCK

Jacob Weimar began as a safe maker for Herring Farrel & Co. at least as early as 1846 and by 1861 he had risen to be the superintendent of Herring's companies.[47] In addition to safe construction and administration, Weimar also designed a combination lock that was available through the Herring Lock Co. beginning around 1869. Weimar's combination lock featured three tumblers, each operated by its own indirect-drive dial by way of three concentrically nested spindles. The bolt handle appears from the outside to be a fourth dial, bearing markings similar to the tumbler dials, possibly to confuse potential thieves. Of the hundreds that are thought to have been made, fewer than ten examples of the Weimar combination lock are known to survive today.

Jacob Weimar's combination lock. Many people whose work was focused primarily around the administration and supervision of safe and lock factories also patented their own lock designs, such as Jacob Weimar. His low-cost three-tumbler combination lock used three separate concentrically nested dials, each with forty discrete positions marked with numbers 0–12 interspersed with letters A through Z as well as an ampersand, visible between Z and 1.

HALL'S SAFE & LOCK CO. REVOLVING BOLT COMBINATION LOCK

Despite the popularity and success of Joseph Hall's Premier combination lock, Hall's Safe & Lock Co. would continue to introduce new combination locks throughout the 1870s. A salesman's sample of one such design is seen here with glass covering a skeletonized back plate, showing the hand-changing four-tumbler mechanism and revolving bolt.[48] Original production figures for this lock are unknown, but today fewer than ten are known.

Hall's Safe & Lock Co. Revolving Bolt combination lock. Lock makers put significant effort into making their exhibition pieces and salesman's samples eye-catching and informative. This sample made by Hall's Safe & Lock Co. in around 1870 features both a standard and a cutaway cover, allowing a prospective buyer to observe the fine four-tumbler mechanism in action. The cover on a production model would be solid, with the concentric fluting running all the way around the central silver portrait of Joseph Hall himself.

HALL'S SAFE & LOCK CO'S. 150-NUMBER COMBINATION LOCK

Introduced possibly as early as 1870 by Hall's Safe & Lock Co., this combination lock is thought to have been developed by Henry Gross. The five-tumbler mechanism featured a 150-number dial that required not only exquisite precision in production but also above-average precision in day-to-day use. Also notable is that the Hall 150 has a key-changing combination mechanism. Key-changing combination locks had been known since 1865, but this seems to have been the only key-changing design ever made by Hall's Safe & Lock Co. This salesman's sample is mounted on a wood block with a marquetry border and may be the only surviving example.

Hall's Safe & Lock 150. Henry Gross was a major figure at Hall's Safe & Lock and was responsible for a number of increases in bank lock security. One effort to improve upon Hall's already popular Premier involved increasing the numbers on the dial from 70 to 150, making a thief's near miss of the combination less likely to open the lock. However, this finely numbered dial required the owner to have an unusually steady hand, since even the smallest overshot of the dialed number would likely require redialing.

DIEBOLD & KIENZLE'S COMBINATION LOCK

Probably first made around 1870, this five-tumbler combination lock is thought to have been made by the Peerless Lock Co. for Diebold & Kienzle,[49] a predecessor to the Diebold Safe & Lock Co. An ornate five-tumbler mechanism, this lock is bronze with surface jeweling throughout and the scarlet interior paint that identified a maker's best lock. Its unique jaw-bolt mechanism opens when unlocked to receive a portion of the boltwork. Of the several hundred that are thought to have been made, fewer than ten examples of this lock are known.

Many bank locks featured intricate surface decoration throughout their construction, but rarely were locks made with sculpted internal ornamentation like that found on this Diebold & Kienzle combination lock. Unique to this lock is the jaw-bolt, visible at the right side of the case. When unlocked, the two sections of the bolt part, allowing a portion of the boltwork from the vault door to slide into the case. The jaw-bolt is a cousin of Sargent's widely successful rollerbolt, as both designs use lateral movement around an axle to prevent bolt pressure from reaching the tumblers.

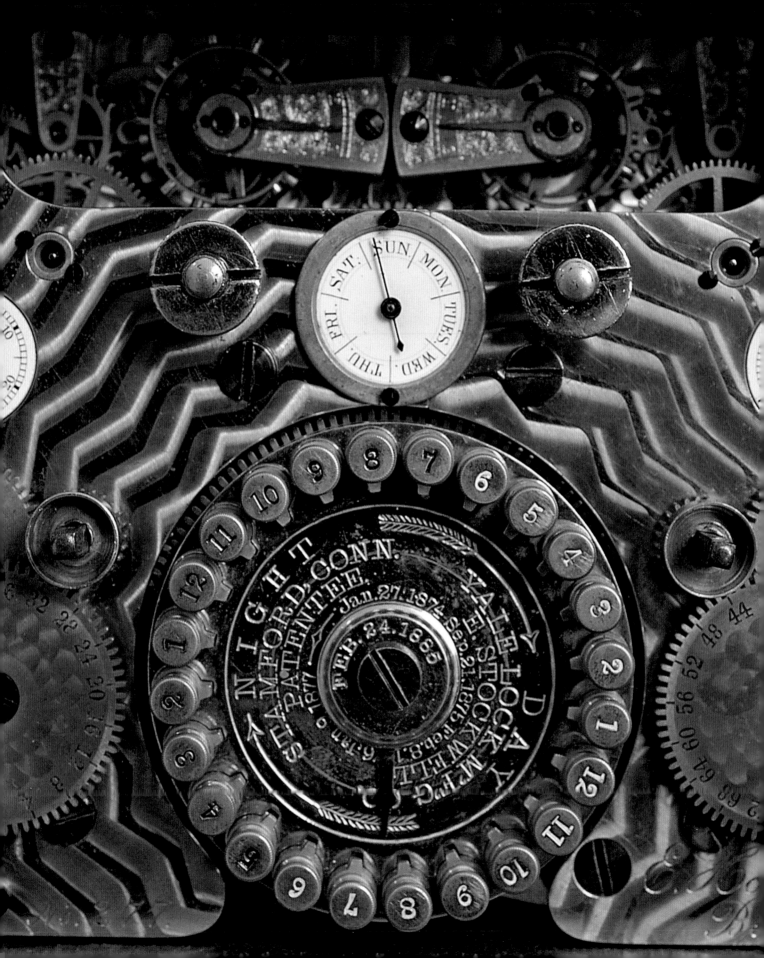

CHAPTER 4

The Rise of the Time Lock: 1872–1888

The use of a clock mechanism to secure a safe did not become common until more than forty years after its first English patent in 1831. Early bankers were justifiably concerned about the reliability of a lock that barred entry by friend and foe alike, since a failed time lock was expensive, requiring the destruction of the safe. But by the 1860s, advances in watch and clock technology had made the mechanisms very reliable while advances in safe and vault construction had frustrated burglars, so much so that they began kidnapping bank employees to obtain combinations. With the beginning of these "masked robberies," as they were called, bankers now embraced the time lock. A period of innovation and improvement began.

JOHN BURGE'S GOTHIC TIME LOCK

In 1872 John Burge, a physician in Circleville, Ohio, designed what has come to be known as the Gothic time lock. Burge became interested in time lock manufacture around 1869 and in 1872 was awarded his first U.S. patent for a design likely too basic and unrefined to be produced and sold. By May of 1872 Burge had developed plans for the more refined Gothic time lock. He brought these designs to Forestville, Connecticut, where he met with prominent clock maker Laporte Hubbell, who constructed the example we see here to Burge's specifications. Hubbell included his patented Marine Clock movement, which uses two independent mainsprings to drive a single platform escapement.[1] The forty-eight-hour movements turn a wagon-wheel-style dial that triggers a bolt-release mechanism powered by its own independently wound spring system. This automatic bolt system rotates the round bolt to align its cutout with the boltwork.[2] The bolt shape is clearly similar to Sargent's unpowered, gravity-dependent rollerbolt, a design first found in Rutherford's 1831 English time lock patent.[3]

Burge received a Canadian patent for his design in 1873,[4] but he did not immediately file for a patent in the United States. Instead, for reasons still unclear, Burge seems to have sought a buyer for the design, marketing it throughout the northeast and mid-Atlantic states to both safe and lock makers and directly to bankers without success.[5] This delay in filing his patent application was probably Burge's undoing.

Burge did eventually sell the Gothic time lock design to Thomas Keating, an agent for Yale Lock Manufacturing Company, in 1875 along with the right to seek a domestic patent and the rights to his 1872 patent.[6] With Yale and Sargent in the opening stages of their vigorous time lock competition, both in the market and in the courtroom, Yale funded Burge's application for a patent on the Gothic time lock in July 1876. With issues surrounding Sargent's 1873 patent still unsettled, Yale was interested in any means of showing patent priority, including the possibility of a late patent on an earlier design. Burge was granted his Gothic time lock patent in August 1877, more than five years after the Gothic time lock was first made.[7]

Though the Gothic time lock was originally intended and used by Burge as a sample, it eventually served as the patent model and was on file with the Patent and Trademark Office during the 1875 hearing before the commissioner. In 2001, watch and clock expert David Christianson completed a painstaking two-year restoration of the Gothic time lock, returning this unique and important piece to full working order.

Thomas Keating. Keating was a major figure at the Yale Lock Mfg. Co., not only negotiating the sale of Burge's Gothic time lock but also working as a Yale salesman from 1869 until 1898. After leaving Yale, Keating joined John Mossman in New York before taking over his operation following Mossman's retirement.

GOTHIC TIME LOCK

OPPOSITE: *John Burge's Gothic time lock. The idea for a timer-operated lock dates to at least 1831, yet despite a number of American patents and the small-scale functional success of the Holbrook Automatic in 1858, there was little enthusiasm among bankers for a lock that made no distinction between its owner and a rank stranger. However, by the time John Burge had famed clock maker Laporte Hubbell construct this fully functional model of his Gothic time lock in the early 1870s, the time lock was poised to become the last and possibly most important major development in bank security for the next one hundred years.*

ABOVE: *Although roughly finished, the Gothic time lock was replete with features, including two independent power springs (winding arbors visible at the bottom of the lower plate), a reliable marine chronometer movement, and a powered rotating bolt. The broad-rimmed central dial indicated the time before unlocking, zero to forty-eight hours. Originally used by Burge as a marketing sample, the Gothic time lock later served as a patent model and, eventually, as evidence in a patent lawsuit between Sargent & Greenleaf and Yale Lock Mfg. Co.*

J. H. SCHRODER & CO.'S BANK LOCK

In the mid-1800s, a number of members of the Schroder family were making locks in and around Ohio.[8] While it is not clear exactly when this lock was produced, J. H. Schroder & Co. of Cincinnati is known to have made bank locks as late as the 1870s.[9]

This extraordinary bank lock features a Bramah-style secondary lock and an eight-lever changeable-bit primary lock. The primary Bramah-style key is inserted and the first quarter-turn retracts the block from the primary keyhole. The secondary key is inserted and the shaft unscrewed, leaving the bit in the keyhole. Then the primary key is turned the last three quarters like a wrench, carrying the bit up to the levers and throwing the bolt.[10] With high-security combination locks well established in the banking industry by 1870, Schroder's bank lock was likely obsolescent when new and may have been the last high-complexity key lock developed in the United States.[11] The scope of Schroder's bank lock production is unknown, but only one other less-sophisticated bank lock made by J. H. Schroder and one unassociated key are known to have survived.[12]

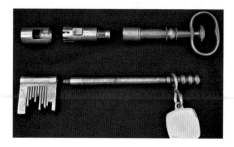

J. H. Schroder's bank lock. While key locks would always remain common in Europe, combination locks were already well on their way to dominating the American bank lock market by 1870. With many American key locks being made for lower-profile and -security purposes, Schroder's bank lock ran contrary to the trend. His two-key solution, though obsolete, carried a noble pedigree: the primary lock is based on Bramah's 1784 design (slider ends clearly visible through the top keyhole) and the secondary lock featured a Mack-style keyhole block, a Lewis Lillie–style removable shaft, and a Solomon Andrews–like bit. Even the lever mechanism looked back to the Day & Newell Parautoptic with its finely toothed catches visible at the bottom right. (Courtesy of Lynn Collins.)

SARGENT & GREENLEAF'S MODEL 1 TIME AND COMBINATION LOCK

In 1874, Sargent & Greenleaf introduced its first time lock, the Model 1. Some of the design details, such as the wagon-wheel dials, suggest that James Sargent was familiar with Burge's unpatented Gothic time lock, but this may also have been simply the two designers' reliance on conventional clock makers' gear shapes. Unlike the earlier time lock designs of Pye, Holbrook, and Burge that operated on a safe's boltwork, the Model 1's time lock acted directly on the combination lock's fence, holding it aloft over the tumblers until the designated time.

Sargent's Model 1 was unquestionably the state of the art when it debuted, but it met with only halting acceptance since it required safe and vault makers to design boltworks with the Model 1 in mind, accommodating both its large space requirement and specific locations for the bolt and spindle. The first Model 1 was sold in San Francisco on June 11, 1874,[13] and though it was available for purchase until 1885 it seems that surprisingly few were made. By 1875 Sargent had made nine or ten,[14] but it is unclear how many were made over the next ten years.[15] Two examples of Sargent's Model 1 survive today. The one shown here was donated by Sargent & Greenleaf, Inc., to the Mossman Collection and is thought to have been a salesman's sample. The patent model for this time lock is part of the Harry Miller Collection at Lockmasters of Nicholasville, Kentucky.

MODEL 1

The Sargent & Greenleaf Model 1 time lock. Sargent's Model 1 debuted in 1874 when the American banking industry was sharply concerned with the "midnight" or "masked" robbery. Despite its obvious merits (integrated first-rate combination and time locks, first-rate maker's reputation, beautiful finish) the Model 1 seems to have had limited use, likely due to its requirement of a specially designed boltwork. Consequently, the Model 1 is not only among the most important time locks but also among the rarest.

ABOVE: *Sargent's Model 1 was the first time lock design to operate directly on the combination lock rather than blocking part of the boltwork. When set, the time lock's central drop lever holds the combination lock fence up, away from the tumblers, preventing the lock from opening. These earliest Sargent time locks had only a forty-six-hour maximum time limit. Here, the mechanism is shown with its time lock open (the drop lever has been rotated slightly counterclockwise, releasing the fence pin) and the combination lock closed.*

SARGENT & GREENLEAF'S MODEL 2 TIME LOCK

Concurrently with the debut of its Model 1 in 1874, Sargent & Greenleaf introduced the first style of the Model 2. A time lock only, the Model 2 operated on the boltwork, allowing it to be used in conjunction with any key or combination lock. In all its types, Sargent & Greenleaf would sell about eighteen hundred Model 2s between 1874 and 1929, making it the earliest time lock to be produced in major quantities. Unlike the majority of smaller time lock makers soon to be operating, Sargent & Greenleaf was a movement maker as well, with the clock-making portion of its Rochester, New York, facility overseen by Lyman Munger, a bench-trained watchmaker. There is evidence that Sargent did purchase certain premade clock parts, but the company would always rely ultimately on the internal design and manufacture of movements.

Like other early styles of the Model 2, the Model 2[1] has black enamel wagon-wheel dials with white numbering. The earliest type of Model 2, or Model 2[1],[16] featured two clock movements, each with two palette jewels. The dials on the Model 2[1] are clearly numbered through forty-eight hours but, with a mainspring capable of storing only forty-six hours of power, the Model 2[1] was sold with a caveat that the movements not be wound past forty-six hours.[17] The dials are fixed to their arbors, which extend completely through the mechanism and are secured between the plates by a sliding sleeve attached with set screws. These fixed-arbor dials were difficult to remove and replace during maintenance and repair.

Of the Model 2[1], only one example of the movement survives, serial numbered 66. The serial number is embossed on the far right bottom edge of the back plate. The case and bolt of the Model 2[1] are thought to be substantially similar in design to the Model 2[2], which soon replaced it.

In late 1874, Sargent & Greenleaf produced its first revision to the Model 2, the Model 2[2], which is the earliest Model 2 for which a complete example survives. The Model 2[2] uses two forty-six-hour movements, each with three

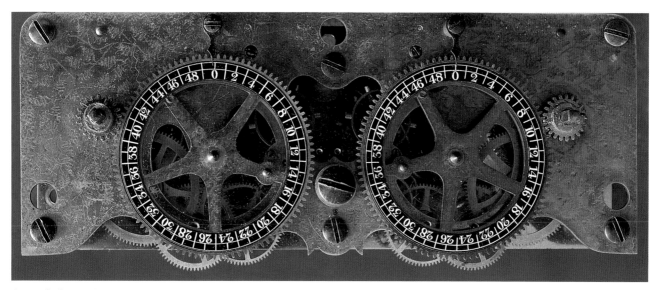

Sargent & Greenleaf introduced its Model 2 time lock for safes and vaults that had sufficient room in the boltwork for a time lock but not enough to use the larger Model 1. No complete examples of this earliest version of the Model 2 are known to exist, but the movement does survive. Like the Model 1, the movement's mainspring could hold only forty-six hours' reserve power. This earliest Model 2 movement featured dials marked through forty-eight hours but came with instructions warning the user not to wind past forty-six hours.

jewels, adding a jewel in the escapement to the two palette jewels of the Type 1. The black wagon-wheel dials retain the fixed-arbor design, but have revised, accurate numbering through forty-six hours. The movement's front plate is engraved with the same leaf-and-vine design found on internal parts of Sargent's Magnetic and Automatic locks. The hook lever, back plate, and rollerbolt screw cover are nickel-plated with a vine-and-star engraving/embossing pattern. The case interior is painted scarlet, in keeping with the lock makers' convention for highest-security models.

The bolt mechanism is a rollerbolt like that first introduced in Sargent's Magnetic lock, a rotating cylindrical bolt with an armature held by the central hook lever and an opening that aligns with an opening in the case when the bolt is released. So long as the hook lever holds the bolt armature up, the bolt-

work will be dogged by a piece of the boltwork that must enter the case. The Model 2[2] rollerbolt is gold-plated bronze with surface "jeweling," giving the frost-on-glass appearance. The similarity between Sargent's Model 2[2] and Burge's Gothic time lock is relatively clear, with the major differences being Sargent's use of two escapements and a gravity-operated bolt.

One drawback of the rollerbolt design inherent in its adaptation from a combination lock to a time lock was that it required an "auxiliary bolt," a device patented by Sargent & Greenleaf in March of 1875.[18] To lock the safe, the owner would (a) lock the auxiliary bolt open by lifting and sliding the knob away from the case, (b) wind the time lock, (c) lift and hook the rollerbolt armature to the hook lever, (d) close the door, and (e) throw the safe door bolt shut. With this last step, the sleeve of the auxiliary bolt, mounted to the moving boltwork, is carried toward a fixed part of the door. As the round, flat backside of the auxiliary bolt abuts the fixed piece, the sleeve, moving with the bolt, slides along the now stationary internal piece, eventually letting it snap shut. Then the boltwork cannot return to its open position without the auxiliary bolt reentering the rollerbolt's opening in the time lock. The only known complete Sargent & Greenleaf Model 2[2] features a solid bronze door, an option for vaults where coinage was frequently moved. The Model 2[2] was produced until late 1875. The auxiliary bolt shown on page 152 is one of two known.

OPPOSITE: *The earliest known complete Sargent & Greenleaf Model 2, this Type 2 was made in late 1874 with a solid bronze door for use in a coin vault where a glass insert would likely get broken. Notably, the dial numbering was revised to reflect the correct forty-six-hour maximum time.*

Near the end of 1875, Sargent introduced the third style of its successful Model 2 time lock (page 153). Like the Model 2[2], the Model 2[3] used two Sargent-made three-jewel movements, but replaced the fixed-arbor dials with dials attached at the front by a screw and washer assembly, making accurate servicing easier. The front plate of the forty-six-hour black-dialed movements continues the rough leaf-and-vine engraving. The Model 2[3] also continued the case and rollerbolt jeweling, hook lever, back plate and screw cover design, and scarlet interior paint. As a rollerbolt-based design, the Model 2[3] required an auxiliary bolt. Four examples of the Model 2[3] are known to survive.

Beginning in 1876, Sargent & Greenleaf again modified the Model 2, introducing the Model 2[4] (page 154). Made through the end of 1876 and into 1877, the Model 2[4] included two forty-six-hour black-dial Sargent-made movements that now featured four jewels each. Sargent no longer included the rough leaf-and-vine engraving on the movement's front plate, but added engraved nickel-plating to the rollerbolt. Unchanged were the jeweled-surface bronze case with a scarlet interior, the symmetrically scalloped door glass, and the need for the auxiliary bolt. Between 150 and 200 are thought to have been made, and only three are known to exist today.

During the same period that the Model 2[4] was in production, Sargent & Greenleaf began applying plaques advertising that a certain safe or vault was protected by its time lock. Similar plaques were applied by Yale Lock Mfg. Co. and Consolidated Time Lock Co. Sargent also prevailed

Sargent's rollerbolt was quite well suited for use in a combination lock, but its use in the Model 2 time lock required the addition of this auxiliary bolt, attached not to the door but to a moving part of the boltwork. Closing the safe meant opening the auxiliary bolt (as pictured above, top), winding the time movements, hooking the rollerbolt up (locked), and closing and locking the door. As the boltwork closed, the outer sleeve of the auxiliary bolt would move with it while the flat end of the inner piece would abut a fixed part of the door. The sleeve would slide until the auxiliary bolt snapped shut (as pictured above, bottom), time-locking the door.

Time locks were intended to discourage thieves from kidnapping bank employees for the safe combinations, yet they were of little benefit if the criminals could not be convinced of the time lock's presence in the safe. Consequently, Sargent & Greenleaf along with other time lock makers began making bronze plaques to advertise the use of their time locks.

OPPOSITE: *Sargent & Greenleaf's third style of Model 2 debuted at the end of 1875. A minor revision, its most significant change was the dial mounting method. In Sargent's first two types of Model 2, the dials were permanently fixed to their arbors, making disassembly and adjustment difficult. The Model 2[3] introduced a dial attached by a screw visible at its center. This small change made maintenance much easier and would continue on all subsequent versions of the Model 2.*

MODEL 2

MODEL 2

Along with plaques outside a vault, safe makers included advertising plaques inside the safe door as well, reminding the buyer that a Sargent & Greenleaf Model 2 could be added later. These time lock advertising plaques are, themselves, quite rare—sometimes less common than the time locks they advertised. Only a handful of plaques are known to survive today.

Probably the most reproduced image of any early time lock, this photo of what is now thought to be Sargent & Greenleaf's Model 2[5] first appeared in a 1927 Sargent & Greenleaf catalogue as an illustration to a historical note. That unfortunately incorrect historical note, claiming this to be the "first time lock ever purchased by a banker," has been the source of much confusion ever since. No examples of this lock are known to survive, but the screw-set dials clearly place this time lock after the Model 2[2] and the redesigned hinge suggests that it was no earlier than the Model 2[4]. Likely a transitional model between the Model 2[4] and the major redesign of the Model 2[6], the Model 2[5] probably had a short production run and limited success.

upon safe makers to include advertising plaques for a short period during the 1890s. These plaques advertised Sargent's Models 2, 3, and 4, and were installed in a space in the door or boltwork set aside by the maker for that Sargent & Greenleaf time lock. Notice and spacing plaques can be even rarer than the locks they advertised.

Possibly the best known and least understood version of Sargent & Greenleaf's Model 2 is thought to be the Model 2[5], probably introduced in 1877. The photo shown here—the only known image of the Model 2[5]—appears in *Lure of the Lock*[19] and seems to originate from a 1927 Sargent & Greenleaf catalogue that offered the following explanation from a letter from the First National Bank of Morrison in Illinois:

> The first time lock ever purchased by a banker was sold to J. J. Jackson, Cashier of the First National Bank of Morrison by Mr. James Sargent of the Sargent and Greenleaf Company, Rochester, N.Y.—This lock was attached to the door of a large iron safe in May, 1874—under the personal direction of Mr. James Sargent.—The lock gave excellent satisfaction and in a very short time there was an urgent demand for similar locks from bankers all over the United States.[20]

This notation seems to be the primary seed that gave rise to the oft-cited but incorrect identification of the Model 2[5] as the "first time lock installed,"[21] when in fact Holbrook's Automatic deserves that accolade. And while we may reasonably forgive the anonymous author for dubbing this the first commercially installed time lock (writing, as he was, more than fifty years after the fact), less understandable is Sargent & Greenleaf's pairing of this quip with a demonstrably incongruous time lock. One can only imagine that the staff of (by then) Sargent & Greenleaf, Inc., had no recollection of the facts and simply photographed the earliest time lock easily available. Interestingly, even the nature of the photo is somewhat unusual because, inexplicably, the rollerbolt has been disabled by removing the rollerbolt armature, clearly missing

OPPOSITE: *By 1876 the Model 2 was clearly a commercial success and Sargent began to address the issue of balancing production and maintenance costs. Changes in the Model 2[4] included additional palette jewels in the movement and reduced engraving on the movement. Introduced in 1876 and made for more than a year, the Model 2[4] was the first time lock to get an extensive production run, thought to be between one hundred and fifty and two hundred.*

MODEL 2

ABOVE AND OPPOSITE: *The Model 2[6] was Sargent's first time lock to address the problems of the rollerbolt and auxiliary bolt, replacing it with the far superior cello-bolt. The cello-bolt is a two-piece design, with the front handle portion made of bronze and the actual bolt portion in the rear with nickel plating. The large pivot point at the left of the handle allows the handle to be lifted and locked onto the drop lever while the door's boltwork extension is still inside the case. When the door is closed, the boltwork extension withdraws and the rear part of the cello-bolt is lifted up by an internal leaf spring, blocking the boltwork. This two-part leaf-sprung design is simple, secure, and inexpensive and would be modified many times, surviving well into the twentieth century.*

in the photograph. Despite the well-known photo, there is no known example of Sargent's Model 2[5].

The only difference between the Models 2[4] and 2[5] that can be stated with certainty is the redesign of the door hinge, replacing the two large exterior hinges with a single, long interior hinge. With production of the Model 2[4] continuing into early 1877, the award of the patents for what would be the Model 2[6] coming in 1877,[22] and the beginning of Model 2[6] production in late 1877, the Model 2[5] was likely a short-lived transitional design made in limited numbers. With no known surviving examples of the

Model 2[5], little can be said with confidence about this design.

After three years of successful sales, Sargent & Greenleaf debuted a major overhaul of the design of its Model 2 in November 1877. The Model 2[6] retained the two four-jewel forty-six-hour black-dial movements, bronze case with surface jeweling, and scarlet interior. Minor changes were made to the door glass, which now had an asymmetric shape, and to the hinges, which were now smaller and external. But in a major change, Sargent replaced the rollerbolt with a new "cello-bolt" design. A far superior solution, the cello-bolt's two-piece spring-loaded mechanism allowed the Model 2[6] to do away with the auxiliary bolt for the first time. The operator now need only wind the time lock, latch the bronze forepiece of the bolt to the hook lever, and close the door, the two-piece leaf-sprung design allowing the cello-bolt's forepiece to be latched while leaving the nickel-plated aftpiece below the boltwork extension. When the boltwork is closed, the extension exits the case and the spring between the two cello-bolt parts lifts the aftpiece, blocking the case opening.

Sargent is known to have started etching its Model 2 door glass at least as early as 1878. Earlier models may have had etched glass, but there is no record or example of it and two pre-1878 examples are known to have original unetched glass. The layout of Sargent's etching would evolve over time but always read "Sargent & Greenleaf" and "Rochester, New York." Later changes in Sargent's corporate structure would be reflected in this glass etching (e.g., the 1896 addition of "Co."). Alternatively, some original door glass was etched "from J. M. Mossman, New York" when sold through Mossman, a major distributor. Etched door glass would remain standard for many time lock makers until about 1910.

The Model 2[6] was Sargent & Greenleaf's first after its initial patent interference victory and was the first to feature patent dates embossed on the bolt. Ultimately the Model 2[6] was made for less than a full year with production numbering about two hundred. Today, three are known to survive.

Between 1878 and 1924, Sargent & Greenleaf made no fewer than ten distinct versions of the Model 2.

The Model 2[7], introduced in 1878, used white enamel dials for the first time. Two examples of the Model 2[7] survive today.

The Model 2[8], first made at the end of 1878, added what looked like a counterbalance to the drop lever, extending it toward the top of the case with a circular end, but this added length seems to be cosmetic. Two of the Model 2[8] are known.

Also around 1878 Sargent introduced the sixty-six-hour Model 2[9] to compete with the recently released Yale Double Pin Dial with a "Sunday Attachment." Although the Model 2[9] did not offer the same level of intricacy as the Yale, the sixty-six-hour mechanisms did, for the first time, allow Model 2 users to lock a vault from Friday afternoon until Monday morning. The Model 2[9] was priced the same as the Yale model, at $450. Two are known to survive.

The Model 2[10], first made in 1880, was the first Model 2 to feature Geneva stops, which guaranteed that there was always sufficient spring power to open the lock. The Model 2[10] also raised the base movement time from forty-six to forty-eight hours. Fewer than five examples of the Model 2[10] remain.

By the end of 1886 seventy-two-hour movements had become the industry standard for time locks and Sargent again updated the Model 2 to the Model 2[11], now with a seventy-two-hour mechanism, replacing both the forty-eight-hour Model 2[10] and the sixty-six-hour Model 2[9]. Notably, the white enamel dials are marked only through seventy hours. The Model 2[11] retained the larger drop lever, but added a small hinge to it, possibly as an anti-dynamite precaution. Fewer than five examples of the Model 2[11] are known to survive today.

Sargent & Greenleaf's Model 2[12] debuted in 1889. The two seventy-two-hour movements now have corresponding dial numbering and the patent dates, though still embossed on the cello-bolt, are in a simple rectangle rather than the Stars-and-stripes pattern of earlier models. The Model 2[12] would

LEFT: *Letter of engagement signed by John Mossman for Sargent & Greenleaf. The engraved letterhead shows a Model 2[7] time lock at the top right and an image of what may have been a preproduction design with a Seth Thomas movement at the top left. No examples of this design with its skeletonized movement are known to exist.*

OPPOSITE: *Herring & Co. bank vault safe. This illustration shows how a Sargent & Greenleaf Model 2[12] would be installed in a Herring Champion safe. This extraordinary safe featured two Herring Dexter combination locks, visible to the right of the time lock. An additional pair of Dexter locks guard the interior door. (From a Herring & Co. catalogue, 1889.)*

MODEL 2

MODEL 2

ABOVE AND OPPOSITE: *Sargent & Greenleaf made many changes and improvements to its Model 2 over the more than fifty years it was in production. By 1889 the Model 2 was a well-known staple of the banking industry and the twelfth version retained a number of improvements introduced in earlier versions, including seventy-two-hour movements, Geneva stops, and the cello-bolt. The Model 2[12] would continue unchanged for seven years, making it the longest-lived variation. Still, fewer than ten examples are known to survive today.*

be unchanged for seven years, but fewer than ten of this type are known today. The retail price of the Model 2[12] was $400 from 1874 into the 1890s, the same price as the standard Yale Double Pin Dial.

Sargent & Greenleaf had always been a partnership, but in 1896 Sargent & Greenleaf incorporated, becoming Sargent & Greenleaf Co.[23] The change was reflected in the Model 2[13], which included "Sargent & Greenleaf Co." on the dials. Other changes included the use of unscalloped rectangular door glass and the replacement of the lever lock for the door with a simple, inexpensive handcuff-style lock. Made until about 1910, the Model 2[13] was the most numerous style when made, and is today the most common, with as many as fifty examples thought to survive.

Around 1910, Sargent made what seem to be primarily cost-saving changes, introducing the Model 2[14], the first Model 2 with no engraving at all and the first without patent dates since the 1877 advent of the cello-bolt. After 1910, all

engraving and patent dates would stop forever across all Sargent's time lock lines. By this time, Model 2 production was already beginning to slow and fewer than ten examples of the Model 2[14] are thought to survive today.

The next significant change to the Model 2 was also precipitated by a change in Sargent & Greenleaf's structure. In 1918, Sargent & Greenleaf Co. reorganized, and from this time on, its products would be marked "Sargent & Greenleaf, Inc." The Model 2[15] shows this change on the dials.

After 1924 all Model 2 time locks were made with a satin nickel finish, although the raw bronze finish was still available as a custom order.[24] These very last of the Model 2s seem to have been sold primarily in Canada and possibly overseas,[25] and though there are most likely some few of these that survive, no complete examples are currently known, making it impossible to say for certain whether these twilight numbers deserve their own sixteenth type.

According to surviving shipment records, Sargent & Greenleaf, Inc., discontinued production of the Model 2 soon after 1927, a year when Sargent was making fully twelve different styles of time locks in eighteen sizes. Not surprisingly, the beginning of the Great Depression put a halt to new bank construction and, by extension, to almost all new time lock production.

YALE LOCK MFG. CO.'S DOUBLE PIN DIAL TIME LOCK

In 1875, the Yale Lock Manufacturing Co. introduced a time lock to meet the then clearly growing demand for time locks, specifically Sargent & Greenleaf's Model 2. Although this time lock is engraved with patents attributed to Little and Burge, it is primarily based on a patent by Emery Stockwell.[26] Stockwell was a dominant figure in the time lock industry for twenty years. He began his career working with Linus Yale, Jr., in Shelburne Falls in 1869, eventually becoming the superintendent of the bank lock department, where he remained until his death in 1891.

The Double Pin Dial time lock originally featured two forty-eight-hour movements made by Seth Thomas, as illustrated in a Yale flyer from 1875. But the Seth Thomas movements proved too unreliable, and in June of 1876[27] Yale

Close on the heels of Sargent's 1874 introduction of the Models 1 and 2, the Yale Lock Mfg. Co. debuted its own time lock, the Yale Double Pin Dial time lock. A veritable Rube Goldberg mechanism, the Double Pin Dial had many innovative features, but at the cost of being one of the most complicated time locks ever. The Double Pin Dial could automatically lock or unlock for each hour of the day depending on how the user set the pins and, with the optional Sunday Attachment, those settings would be automatically skipped once every seven days. It was significantly more complicated and expensive to produce than Sargent's Model 2, though both sold for $400 retail. The Double Pin Dial shown here was made later, after the 1883 name change to Yale & Towne Mfg. Co., and features the Sunday Attachment, a $50 option.

Emery Stockwell. Stockwell began his career in the lock industry with Linus Yale, Jr., in 1869 and worked for Yale Lock Mfg. Co. until his death in 1891. Stockwell was awarded numerous patents and was instrumental in developing many of Yale's most important time lock designs.

switched to a fifty-six-hour movement made by E. Howard.[28] There are no known surviving examples of the first forty-eight-hour Pin Dial, and it is thought that all the Seth Thomas movements were rapidly replaced, through either attrition or an organized campaign by Yale.

The Pin Dial, clearly a far more complicated design than Sargent & Greenleaf's Model 2, offered a number of innovations, the most obvious being the two front dials, each with a series of twenty-four pins. Every pin, embossed with an hour of the day, could be individually pulled out, opening the door for that hour, or pushed in, locking it, making the Pin Dial the first time lock capable of both unlocking and locking a safe at a predetermined time. Further, the Pin Dial was the only mechanical time lock ever to be able to lock, unlock, and relock a safe for various one-hour periods throughout the day.

In 1876 Yale added the Sunday Attachment, which allowed the lock to skip the opening periods each Sunday. The mechanism is identified by the two curved nickel runners next to the inner top third of the pin dials embossed with "thursday friday saturday" and a small shield on each dial that is geared to rotate at one-seventh the rate of the dial. These shields move over the names of the days, allowing the user to check to make sure it is accurate, and then, during the twelve hours of Sunday day, will be interposed between the drop wheel and the pins, guarding the openings of the pulled pins and keeping the lock shut.

It is not clear exactly when the Pin Dial's standard movement changed from fifty-six to seventy-two hours. An analysis of serial numbers and production ledger dates suggests a change as early as 1879, but catalogues advertising the availability of seventy-two-hour movements from Yale would place the change much later, in the mid- to late 1880s. Many

ABOVE AND OPPOSITE: *One of Yale's earliest time lock advertisements. This advertising flyer from 1875 is the only known record of how Yale's first time lock design using Seth Thomas movements may have looked.*

fifty-six-hour movements were rebuilt into seventy-two-hour movements, a service offered by E. Howard for $30.[29]

The Yale Pin Dial cost $400 ($450 with the Sunday Attachment) during the entire period of its sale, from 1875 to 1900,[30] although actual production tapered off and had substantially stopped as early as 1892.[31] In the first four years Yale made thirteen hundred Pin Dials, but the company made only slightly more than eight hundred more between 1879 and 1892.[32] An order for seventy-seven Pin Dial movements numbered 2153 to 2230 at $54.25 each appears in the Howard ledger for the year 1889, and a last order of seventy-five at $55 appears for 1890.[33] This 1890 order in Howard's ledgers does not specify the serial numbers of the last known batch of Pin Dial movements, but sequential numbering would run through 2305. There is some evidence that Pin Dial production and sales continued even after Howard's recorded production ended, with some later-numbered movements known, almost exclusively overseas, primarily in England. The Pin Dial is a relatively common time lock, with some four hundred or more surviving, among which about fifty have Sunday Attachments.

HALL'S SAFE & LOCK CO.'S TIME LOCK

According to the testimony of Joseph Hall in the case of *Yale v. Berkshire Nat'l Bank*, his first time lock arrived from E. Howard on July 1, 1875.[34] It was made to the specifications of a design by Henry Gross, who described himself as "the inventor" at Hall's Safe & Lock Co., according to his testimony in *Hall v. MacNeale & Urban*.[35] Gross received a patent for this design in February of 1876.[36] Even in this earliest model, Hall's time lock was designed to operate directly on a combination lock, a significant departure from the then successful designs of Sargent & Greenleaf and Yale that operated only on the boltwork.

The overall design of the forty-eight-hour movements resembles a carriage clock having platform escapements viewable through the case top, a style of clock popular at the time. This early design has a number of notable features. The first movement design featured a comma-shaped escapement wheel bridge with the two movements mounted in the case on a single shared plate attached to the case back with four small screws set between the edge of the movement and the side of the case. The case is silver-plated bronze. Production models would use a simplified shield-shaped bridge and mounting screws through the case back.

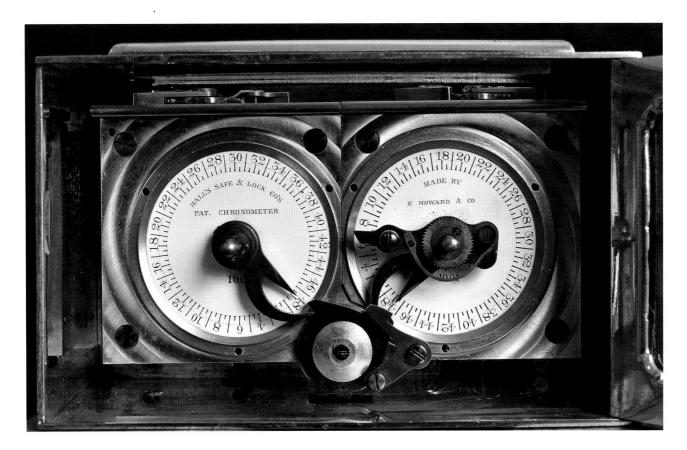

Hall introduced this time lock around the same time Yale introduced the Double Pin Dial, and wholesale production prices of the two were similar, according to production ledgers of the E. Howard Watch & Clock Co.,[37] with the Hall movements' price set at $17.50 each, according to the earliest production reference on July 16, 1877.[38] A small number of these earliest working models are thought to have been delivered to Hall. One such model, serial numbered 1002, was pictured in Hall's 1880 catalogue and was later an evidentiary model in lawsuits. The example shown here is serial number 1001.

Hall's Safe & Lock Co. of Cincinnati was one of the few major bank lock makers outside the American Northeast in the 1870s. With the company's market centered around Chicago and the West, Hall's Safe & Lock patented its first time lock in 1876 for banks that relied on security to prevent theft rather than law enforcement to solve it. Its first model had many features that would become common in Hall time locks, such as its hook release through the case bottom designed to operate directly on a Hall combination lock. The single large, rectangular top window would eventually be replaced by two smaller, stronger round windows.

Hall's serial numbering of time lock movements began at 1001, but did not proceed sequentially. For example, a February 1890 order requested that one hundred single-movement time locks be numbered 3676–3699 and 4926–5000.[39] Rather than sequential numbering, Hall seems to have set aside blocks

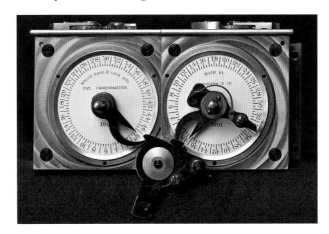

Hall's Safe & Lock Co. time lock movement. This movement was made by E. Howard of Boston for Hall and is serial numbered 1001, marking it as the first movement delivered to Hall's Safe & Lock Co. in July of 1875. Hall would go on to use an intricate serial-numbering method for his time locks that is still not fully understood. However, detailed production records from E. Howard clearly show this to be the earliest Hall time lock movement.

of numbers for particular styles. Double-movement time locks marked "Joseph Hall" seem to be numbered from 1001; single-movement "Joseph Hall" time locks from 2001; single-movement time locks marked "Consolidated" from 2501; double-movement "Consolidated" time locks from 3001; Consolidated time locks with movements labeled from the Harvard Clock Co. are known to be numbered over 3600, possibly starting around 3500. The Harvard Clock Co. was founded in October 1880 and became the Boston Clock Co. in May 1884. After Boston Clock failed in 1894, Joseph Eastman (later of Eastman-Kodak) tried to revive the company in 1895, but creditors foreclosed on the firm in 1896. Some unlabeled movements numbered in the 6000s are thought to have come from Boston Clock for their escapement and movement structure are clearly not from E. Howard.[40] As a consequence, movement serial numbers are only partially helpful in determining the production date of time locks made by Hall or Consolidated. Based on an analysis of Hall's nonsequential serial numbering scheme, it is suspected that two hundred to three hundred of these earliest double-movement time locks were made between 1876 to 1881; fewer than ten are known to exist.

Although Joseph Hall formed the Consolidated Time Lock Co. in January 1880 to insulate his successful safe and lock business from his risky and untested time lock business, the name Consolidated Time Lock Company did not appear anywhere on a time lock[41] (only "Joseph L. Hall") prior to 1882. Around this time, Consolidated stopped including "E. Howard" on the enamel dials, replacing it with Gothic typeface embossing on the face plate behind the dial to obscure it from view. This may have been an attempt by Howard to avoid further litigation after a Yale lawsuit[42] for using a Holmes time lock in its safe. Such lawsuits may also have been a factor in Consolidated's use of frosted door glass.

OPPOSITE: *Many time locks were streamlined between initial design and large production. Some changes were small, such as the difference between the first comma-shaped escapement wheel bridge and the later shield-shaped wheel bridge.*

YALE LOCK MFG. CO.'S STOCKWELL DEMONSTRATION TIME LOCK

Even after Yale Lock Mfg. Co.'s purchase of the rights to the Gothic time lock design from Burge in 1875, Yale still needed to demonstrate that the Gothic could work in an actual bank vault for it to be useful in its patent actions. Emery Stockwell had two models of the Gothic time lock made and installed one free of charge in the Stamford National Bank of Stamford, Connecticut, where it remained in operation for over six months.[43] Although no period pictures of these two models are known to exist, Stockwell gave detailed descriptions in a patent interference deposition[44] that match this model, found in the collection of an antiques dealer in 1998.

This time lock was made by Seth Thomas using an eight-day marine clock movement similar to the Hubbell movement in Burge's Gothic time lock. Notably different is the twelve-hour clock face that replaces Burge's forty-eight-hour wagon-wheel dial. The case is made of wood in the fashion of Seth Thomas clocks with the addition of mounting holes through the exterior flanges of the case bottom. A strong spring-driven automatic is released in much the same manner as the Burge time lock's automatic. The model worked at the time of the trial in 1876[45] and still works today. It is pictured here with a reproduced actuator arm based on concurrent Yale designs and the cutout in the case wall. This is the only surviving example of the two models made.

When the Yale Lock Mfg. Co. installed two time locks based on Burge's Gothic time lock in 1875, time lock design was not yet well established. Unlike other early styles, Yale's 1875 Stockwell time lock had much more in common with clocks of the day than with bank locks, with a wooden case and large twelve-hour face. The mechanism worked well, however, even offering a relatively strong automatic release that could pull open a moderately sized safe bolt when connected by the projecting armature.

STOCKWELL DEMONSTRATION TIME LOCK

HERRING & CO.'S INFALLIBLE TIME LOCK

This double-dial Dexter combination lock was retrofitted with the Herring Infallible time lock. The Herring Infallible was based on a design by Charles O. Yale[46] with improvements developed by Jacob Weimar.[47] In an almost incredible historical exacta, both Hall's Safe & Lock Co. and Herring & Co. were simultaneously developing single-movement time locks that relied on secondary combinations and would be dubbed "Infallible." Herring's Infallible was based on a small single clock movement (likely forty-eight hours), of modest quality compared to those of their contemporaries (*viz.*, Waltham and Howard), and seems to have been made solely as an upgrade to the Dexter combination lock as a replacement back plate.

In the event the Herring movement failed, the user would first dial one combination, leaving the dial on the last number, and then dial the second combination. Then that second dial would be worked back and forth between the last number and the opening point once per second—but no faster—each motion advancing the time lock.[48] The result of Herring's design was that, even if a potential thief knew both the combination and the opening procedure, the lock would have to remain shut for at least the duration set when locking, as opposed to Hall's Infallible, which could be opened immediately after a telegram to Hall's Cincinnati offices.

As was common among lock manufacturers, Herring seems to have begun producing its Infallible prior to the patent award. This led to what was possibly the first time lock patent lawsuit, filed by Sargent & Greenleaf against the National Mohawk Valley Bank in December of 1875 immediately after receiving a patent reissue.[49] The defendant's answer articulated a number of defenses, including that the underlying patent reissue was void because it expanded the rights under the original patent, the very defense eventually upheld by the Supreme Court in its landmark 1890 decision in *Yale Lock Mfg. Co. v. Berkshire Nat'l Bank*.[50] Unfortunately, Herring & Co.—like so many other time lock makers—had neither the money nor the inclination to pursue protracted litigation after two years of depositions and briefs looked likely to lead to Herring's defeat. Production of the Herring Infallible ended in 1877, with only a handful thought to have been made, two believed to have been installed, and only the example shown here known to survive.

The risk of getting locked out of one's safe when a time lock stopped running was always a paramount concern to bankers and designers alike. The most common solution was to add one or more redundant movements, but this was expensive. Another solution was to build a secondary mechanism into the combination lock that could release the time lock should it stop, as was done in the Herring Infallible time lock, seen here mounted in the cover of a Dexter Double Dial. Herring's Infallible mechanism allowed a user to advance a stopped time lock using one of the combination dials as a type of pendulum, turning it back and forth until the time lock released. The connection can be seen at the center right of the open case.

YALE LOCK MFG. CO.'S TIME AND COMBINATION LOCK

With the debut of Sargent & Greenleaf's Model 1, Sargent's main competitor, Yale Lock Mfg. Co., sought to respond. On October 3, 1876, Yale was awarded patent 182,868 for its own single-case time and combination lock designed to counter—and outstrip—Sargent's Model 1.

The Yale single-case featured two three-tumbler key-changing combination locks compared to Sargent's one and a Yale Double Pin Dial time lock fitted with the earliest known example of Yale's Sunday Attachment. It also featured an innovative anti-micrometer mechanism different from both Sargent's Magnetic and Automatic locks: with a two-piece bolt handle and bolt connection assembly (visible at case right) the combination locks and time lock actually hold the bolt handle disengaged, engaging the boltwork only when unlocked. Until the time lock has run and one combination lock opened, the bolt handle will spin freely, giving no resistance for the micrometer to measure.

While it may have originally been intended as a flagship design, the Yale single-case would yield little ultimate benefit for Yale. This is the only known example of the design and may be the only one ever made. The absence of patent dates or serial numbers embossed on the case or bolt, the unusual gold-plated parts of the combination lock, and the combination locks being fitted with finger screws rather than spindles and dials all suggest that this example was either the patent model or an exhibition sample.

There is no known record that the Yale single-case was ever produced or installed and it does not appear in any known Yale catalogue, which may have been due to a number of factors. This design would likely have carried an almost prohibitive price tag, likely in excess of $600. Further, as with Sargent's Model 1, the vault door design had to specifically accommodate the Yale single-case, essentially requiring a bank to first buy the lock and then have the vault door's boltwork built around it. Finally, after the October 1877 agreement between Sargent and Yale to limit competition, this model—designed solely to compete with Sargent's Model 1—was rendered an unnecessary and expensive experiment.

YALE TIME AND COMBINATION LOCK

ABOVE: *As in any market with a small number of major companies, time lock makers kept careful watch on their competitors. In 1876, Yale Lock Mfg. Co. patented its single-case bank lock, primarily because it was not yet clear how successful Sargent & Greenleaf's Model 1 would be. With two high-quality combination locks controlled by a Double Pin Dial time lock with a Sunday Attachment, Yale's single-case was clearly state of the art. Though it was never put into production, it would have suffered from the same difficulties as Sargent's Model 1, including the need for a custom boltwork.*

OPPOSITE: *Yale Lock Mfg. Co. billhead. This elegant billhead from 1878 shows a number of Yale's manufacturing interests and notable prizes. Such prizes were held in high regard and were advertised extensively by intensely competitive companies such as Yale.*

EDWARD J. WOOLLEY'S FLUID TIME LOCK

E. J. Woolley, a self-described "safe salesman" for Joseph Hall,[51] worked for Hall's Safe & Lock Co. in various capacities, but his most unusual contribution was undoubtedly his time lock. Woolley's time lock is made up of a cylindrical case and a mounting bracket,[52] and as shown in the patent illustration the case interior is divided into two parts, with a weight on one side (*W* in Fig. 2 below). The unweighted chamber is filled with liquid, which more than offsets the mass of the weight. As the liquid slowly passes from the left chamber (n) to the right (n'), the weight slowly turns the case in the bracket and aligns an opening with the boltwork. The liquid ducts are controlled by a screw, regulating the flow and, therefore, the speed of the time lock. Of an uncertain small number ever made, only two are known to exist, both in the Mossman Collection. The example shown here was donated by Edwin Holmes and is still mounted on the plate that held it in a bank vault.

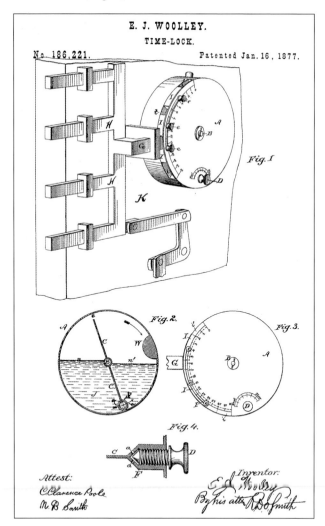

LEFT: *The interior mechanism of Woolley's time lock is best shown by the patent drawing. The two chambers were the same volume, but one included a fixed metal weight (Fig. 2). The weight pulled its side down and the fluid level in the opposite chamber rose, eventually balancing the drum. As the fluid passed through an adjustable valve (Fig. 4) from one chamber to the other, the drum slowly turned as the weight eventually reached the bottom, opening the lock.*

OPPOSITE: *E. J. Woolley's time lock is the only design known to completely do away with a conventional clock or watch mechanism, instead relying on an asymmetrically weighted cylinder that slowly turns as fluid seeps from one chamber to another. A clock based on dripping water was an ancient idea, appearing prior to 1000 B.C.E. in India, China, Greece, and other areas, but such clocks were inaccurate over long periods due to water's low viscosity. For shorter periods, however, water-based clocks worked well and Woolley's time lock was likely quite good for a time lock with only one moving part.*

LEWIS LILLIE'S MODEL 1 TIME LOCK

Also patented in 1877[53] by Lewis Lillie was his Model 1 time lock. Lillie was a significant maker of safes and locks at this time, though his operation was not equal to that of Sargent, Yale, or Hall and his entry into time locks marked what would become a broad influx of small, independent designers into this newly lucrative field. Lillie's Model 1 time lock was designed around the double-main-spring, single-escapement 8-Day Marine Clock movement made by Seth Thomas. Designed to open at a set time each day and lock ten hours later, Lillie's Model 1 was the first to offer what would come to be known as a "calendar" mechanism, a setting that skips the scheduled opening on any given day. The calendar mechanism was similar to Yale's Sunday Attachment, but could be selected for the following day at any point, whereas the Sunday Attachment operated automatically every seventh day. This calendar mechanism would eventually be adopted by a number of makers, including Stewart and Mosler. There are no dependable records of the scope of Lillie's production of his Model 1 time lock, but few are thought to have been made. After Yale sued Lillie for patent infringement, Lillie assigned his patents to Yale. The three surviving examples all differ slightly, suggesting that the Model 1 may never have reached a final design before its 1878 replacement by Lillie's Model 2.

LILLIE'S MODEL 1

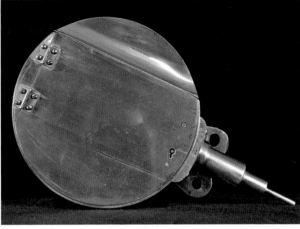

By 1877 time locks had taken the American banking industry by storm and smaller companies began to mobilize, trying to attract some of these new revenues. Lewis Lillie's medium-sized lock company was well positioned to become a second-tier time lock maker and was one of the first to patent an alternative to the Sargent, Yale, and Hall time locks that were quickly becoming the industry standards. While Lillie's Model 1 time lock was technically sufficient and introduced the first true "calendar" mechanism, it was never elegantly finished as Lillie's production was cut short by a patent infringement suit brought by Yale.

SARGENT & GREENLEAF'S MODEL 3 TIME LOCK

In addition to its flagship Model 1 and its commercially successful Model 2, Sargent & Greenleaf began making a Model 3 in 1877. Smaller than the Model 2, the Model 3 used a shorter and shallower version of the newly introduced cello-bolt, allowing it to fit in a host of smaller safes and vault doors with tight boltworks. The earliest Model 3 had a half-solid door, vine engraving on the movements and bolt, and debuted with the black dials and counterbalanced drop lever that were introduced on the Model 2[8]. The square, engraved bushing that guides the boltwork into the case is adjustable, using two set screws. This earliest version displayed no patent dates, since Sargent had yet to win its watershed 1877 administrative patent decision before the secretary of the interior.

Later production of the Model 3 is identified not only by the appearance of patent dates, but also by certain design changes that seem to be in response to Newbury's research on the effects of dynamite on time locks between 1881 and 1884: the clearance between the dials and drop lever was increased by 3/16 of an inch, a third heavy supporting screw was added to hold the movement in place, and mounting projections on

MODEL 3

Whether or not an existing safe or vault could be fitted with a time lock depended on available space in the boltwork, a limitation that often meant that both Sargent & Greenleaf's Models 1 and 2 were too large. For smaller safes and those with tighter boltworks, Sargent introduced the Model 3. Smaller, but equally secure, the Model 3 was made possible by its adapted form of the cello-bolt, which was concurrently introduced on the Model 2.

RIGHT: The earliest Model 3 featured a square adjustable bushing held in place by two set screws to guide the door's boltwork into the case. Later versions used a round bushing that did not need manual adjustment, but rather offered some play to prevent binding on the boltwork. This later version also incorporated antidynamite mounts with rubber inserts to absorb shock.

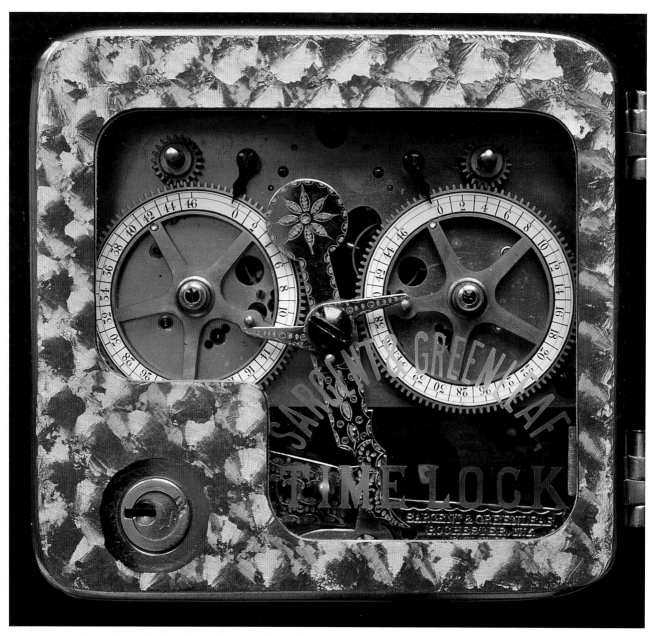

ABOVE AND OPPOSITE: *The use of nitroglycerin and dynamite against safes was of particular concern to time lock makers, since explosive shock could not only jar the time lock open but also damage or stop the movements. The major research in this area was done by Newbury, who filed a series of nine antiexplosive patents in the early 1880s. The later versions of Sargent's Model 3 incorporated many of these improvements, some almost undetectable (such as increased space between important parts), but Newburys patents prevented competing companies from instituting them. Even with these improvements, however, neither Sargent nor any other maker ever claimed its time lock to be explosive-proof.*

the case back were fitted with rubber washers to absorb concussion. By this point the Model 3 featured a round "floating" case-side bushing that allowed for play in the boltwork extension as it enters the case. The Model 3 continued in production until at least 1921.[54] This excellent example was provided by Dana Blackwell from his personal collection.

MODEL 3

LEWIS LILLIE'S MODEL 2 TIME LOCK

By late 1876 Lewis Lillie was making his Model 1 time lock and also was part of the myriad patent interference claims against Sargent & Greenleaf before the secretary of the interior. However, before Sargent would prevail on these patent issues in 1878, Lillie was sued for patent infringement directly by the Yale Lock Mfg. Co. Lillie and Yale reached a settlement in 1876 under which Lillie assigned all his rights to the Lillie Model 1 to Yale, a move that seemed a total victory for Yale.

However, Lewis Lillie was not to be completely run out of the time lock market, and in 1878 a patent was awarded to S. M. Lillie,[55] thought to be a relative of Lewis, for an improved Model 2 time lock. This Lillie Model 2 maintained much of the Model 1, with its Seth Thomas Marine Clock movement and calendar mechanism. Having S. M. Lillie patent the Model 2 and having it constructed by Loriot & Ostrom[56] (makers otherwise unknown in the lock business) may have been moves designed to skirt Lewis Lillie's settlement agreement with Yale. Lillie's intent may have been to sell his Model 2 to Yale as well, but it is not known what became of this design. Shown here is thought to be the only example of the Lillie Model 2 time lock ever made.

ABOVE AND OPPOSITE: *Lewis Lillie's second style of time lock appeared in 1878, despite a prior legal settlement between Lillie and Yale that ended Lillie's time lock production. Although the time lock was produced by Loriot & Ostrom (probably to comply with the letter—if not the spirit—of the settlement), the design clearly draws upon Lillie's original model.*

NEW BRITAIN BANK LOCK CO.'S PILLARD TIME LOCK

By 1877 the time locks of Sargent & Greenleaf, Yale, and Hall's Safe & Lock were clearly an unmitigated commercial success and were the standard of vault security. A few smaller lock makers that produced equally high-quality combination and key bank locks on a smaller scale designed their own time locks in a bid to gain a portion of this newly lucrative area. Frederic North's New Britain Bank Lock Co. was a small but quite successful bank lock maker in and around New Britain, Connecticut. Well known for its Pillard combination lock and the Isham Permutation and Key Register locks, New Britain Bank Lock introduced a Pillard-designed time lock in 1875. Based on two forty-eight- or fifty-six-hour high-quality unjeweled clock movements made by Laporte Hubbell, the Pillard time lock was the first to feature power reserve indicators that would prevent the lock from closing with less than twenty-four hours' power remaining.[57] Such a power reserve indicator (albeit without low-power antilocking) wouldn't be seen again until Yale's B, C, D, and E models of 1888. The Pillard had a heavy bronze case, bevelled door glass, a twenty-four-hour central dial, and floral engraving on the interior front plate and release mechanism.

Operating the Pillard was not a simple process. Beginning with an open safe, one winds and sets the time lock. Then *with the door still open* the boltwork is thrown shut, the power reserve catch on the side of the time lock is raised until it holds, the boltwork is reopened, the safe door closed, and the boltwork thrown shut. If the minimum twenty-four-hour power reserve is met, the reserve catch falls to rest on the release slide, blocking the bolt. Since the safe might seem locked at this point but is not time locked due to insufficient winding, some Pillards have a notice embossed on the door to wind the lock every day.

Altogether, the New Britain Bank Lock Co. installed at least 170 Pillard time locks throughout the Northeast.[58] However, the early success of the Pillard made it the target of patent infringement lawsuits between Yale Lock Mfg. Co. and the City National Bank of Bridgeport in 1877. Prior to this, many makers had issued letters and flyers claiming patent rights and identifying other time lock makers as selling infringing designs, but none had ever brought suit. In response, makers commonly offered to repurchase their time locks and indemnify banks for any costs, should their time locks be ruled infringing. New Britain Bank Lock was no different. Unfortunately, when the decision of Judge Shipman ruled in favor of Yale and issued a permanent injunction against New Britain Bank Lock from producing its time lock,[59] the firm was forced to reimburse City National Bank for the Pillard time lock. Although this single decision was small, both Yale and Frederic North realized its sweeping consequences: any bank that Yale could locate with a Pillard time lock could be forced to remove the lock. As part of an aggressive campaign, Yale

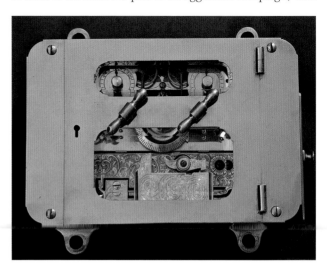

Oliver Pillard's time lock. Much like Lewis Lillie's lock company, Alfred North's New Britain Bank Lock Co. was a successful, moderately sized manufacturer in the Northeast, well positioned to make a bid for part of the time lock market. Unlike Lewis Lillie, however, New Britain Bank Lock's time lock designs by Pillard were an unmitigated success, with more than 170 Pillard time locks installed. Pillard is thought to have introduced three styles substantially simultaneously in 1877, the first one shown here with its two windows to monitor the catch (below) and the movements (above), as well as its fixed winding arbors mounted in the door. The Pillard[1] shown here is late production without any attribution to Pillard, possibly in response to then prevalent patent litigation.

With the door and its mounting panels removed, the full extent of the extraordinary engraving is visible. Though decoration was not unusual on top-quality bank locks and time locks, such deep and intricate scrollwork has not been found on other time locks. This kind of ornamentation must have added significantly to production costs.

offered to sell these banks Yale Double Pin Dials at a reduced price if they surrendered their "infringing" Pillard time locks to Yale. Many banks accepted these terms and Yale promptly melted down the Pillard time locks the company received, making the Pillard among the rarest time locks today. North and New Britain Bank Lock continued for a period but, with New Britain Bank Lock not being incorporated, the mounting losses were passed directly on to North, and he was soon forced to close his doors.

The Pillard time lock was made in three different models, each available with either a polished nickel- or a jeweled gold-plated case finish. The Pillard[i] featured two windows and its pair of winding arbors permanently affixed in the door. The nickel-plated Pillard[i] shown here has the external reserve catch visible on the right of the case. This reserve catch is located inside the case on other examples. Notable on this example are the door glass and the engraving. The door glass is original but shows no engraved attribution, possibly in an attempt to reduce the risk of lawsuits to banks. The unusually elaborate scroll engraving was expensive, even for the time, and is similar to that found on high-grade Parker Brothers shotguns from the same period. Parker shotguns of the late

1870s were made in Merriden, Connecticut, less than ten miles south of New Britain, and it is quite possible that the same craftsman decorated this time lock. This Pillard[i] was found in the Cortland, New York, area and may have been the Pillard time lock known to have been installed in the National Bank of Cortland.[60] One other example of the Pillard[i] with the jeweled gold-plated finish exists in the Mossman Collection.

New Britain Bank Lock introduced two other Pillard designs at substantially the same time as the Pillard[i]. The Pillard[ii] featured a single beveled door window and was wound with a detachable key, like most other time lock designs. In addition to the example shown here with the gold-plated case, one other gold-finished Pillard[ii] survives in the Science Museum, London, and two examples with the polished nickel surface are known, one in the Harry Miller Collection and one in the Lipps Collection, Amsterdam. The Pillard[iii] was a low-profile model designed for use in safes with tighter boltworks. Possibly a special order design, only a single Pillard[iii] with the gold-plated finish remains today as part of the collection at the Lock Museum of America.

PILLARD TIME LOCK

OPPOSITE AND ABOVE: *Like the Pillard[i] time lock, the Pillard[ii] was made with either a nickel-plated finish or the jeweled bronze surface shown here. With only a single piece of door glass, the winding arbors on either side of the central dial were turned with a removable key. All Pillard time lock doors featured a small lever lock, but this did not offer any real security from tampering, since the entire case back could be easily removed with a screwdriver. The early success of the Pillard time locks made them a prime litigation target of Sargent and Yale, litigation that would ultimately completely destroy the New Britain Bank Lock Co.*

HALL'S SAFE & LOCK CO.'S SAMPLE INFALLIBLE TIME LOCK

By the late 1870s it was clear that two high-quality Sargent & Greenleaf or E. Howard time movements were sufficiently reliable for safe and vault owners to trust. But with movements representing the lion's share of a time lock maker's wholesale production costs, Hall introduced his first single-movement time lock in 1878, featuring his Infallible anti-lockout device. The Infallible device depended on a direct connection between the combination lock and the time lock—not an issue for Hall whose time lock designs generally operated on the combination lock rather than on the boltwork. The Infallible device featured an armature from the right-center of the time lock (here, behind the engraved plate below the dial) into the combination lock case behind the bolt. If the timing mechanism failed, the time lock could be released by means of a second combination, derived from the true combination plus a formula. To protect the user from the night-time "masked robbery," this second combination was not known to anyone at the bank but would be cabled to them from Hall's Cincinnati office on request. The mechanism would work only with Hall's Premier combination lock and, though the Premier had been sold since 1869, any Premier could be retrofitted to accept an Infallible mechanism. The Premier was an extremely successful safe lock with many thousands made, and while the Infallible was popular for a time lock, demand for the Premier was so much greater that Hall never modified the Premier case casting to fully accommodate the Infallible's mechanism. Consequently, the single-dial Premier's threaded wheel pack collar would have a patch where the case was cut out to allow for the extra fence, a patch not necessary on the larger double-dial models.

The Infallible time lock shown here is identifiable as one of the earliest with its movement made by an unknown watch

maker and of a style different from all other Hall time locks. It has no platform escapement, but rather a partially skeletonized dial to view the internal, vertical balance wheel. It is unknown why Hall had this movement made, given his sound relationship with E. Howard and the clear inferiority of this mechanism to those that could be expected from Howard. It may have been an attempt to supplement E. Howard–made movements with internally made movements to meet growing time lock demand. Two other examples of Hall time locks with these unattributed movements are known to survive today.

Although these lower-grade movements did appear in production time locks made between 1877 and 1878, this example was probably not intended for installation, but rather for use as an exhibition model or salesman's sample, given the silver portrait medallions on the sides and the engraved panels around the movement. Notably, this example is Hall's first known design that acts solely on the bolt, featuring a bolt-dogging mechanism behind the engraved panel below the dial. It is not known whether this design was ever installed commercially, but it should be noted that this model has no antilockout mechanism should the single movement fail.

Hall's Safe & Lock Co.'s Infallible. The first Infallible time lock by Hall's Safe & Lock Co. appeared in 1878, doing away with the second, redundant time movement in favor of a connecting bar into the Hall Premier combination lock that it would be mounted on. This very early exhibition model of the Infallible has many aesthetic features, including a silver portrait medallion on the side, but includes a low-grade movement from an unknown maker. Later Infallibles would use the excellent E. Howard movements that were common across other Hall time locks.

HENRY NEWBURY'S MODEL TIME LOCK

This time lock is described in *Lure of the Lock* as a time lock model made by Henry Newbury,[61] and though the actual maker of this time lock cannot be stated with confidence, there is good reason to believe that Newbury did indeed assemble this model. Donated to the Mossman Collection by Edwin Holmes in the early 1900s, the model's construction is clearly hand-work using sheet brass and inexpensive time movements available during the 1870s. Newbury was an important inventor and lock technician for Holmes's time lock company during this period and his patents formed a significant part of Holmes's patent litigation defenses. However, an extensive search of time lock patents does not reveal that this design was ever patented, and no mention of the design is made in any Holmes litigation documents. Consequently, though this time lock may have been intended as a patent model, the design is not thought to have ever been patented. An examination of the mechanism reveals that it does not offer any clear innovation that would make it competitive. This Newbury design is the only known example of an unsuccessful time lock seen as a work in progress.

An example of how time lock inventors worked, this preproduction model thought to have been made by Henry Newbury was hand constructed from sheet brass, much like Burge's Gothic time lock. Although the mechanism is complicated, it is similar in operation to other time locks of the period. Without any particular advance in security or functionality, this model was not patented and never went into production, and it is the only known surviving example of a discarded time lock design.

THOMAS F. KEATING'S TIME LOCK

A time lock designed by Yale employee Thomas F. Keating appeared in 1878 featuring a single forty-eight-hour pocket watch movement made by the American Waltham Watch Co.[62] Keating began as a salesman for Yale in 1869. Responsible for Yale's purchase of Burge's patents, he rose to become assistant treasurer by 1882, and in 1906 he took over the time lock business of John Mossman upon Mossman's retirement.[63] The smallest time lock ever made, the Keating time lock operates directly on any combination lock but seems to have been designed with a Linus Yale, Jr., combination lock in mind. The single movement has no backup mechanism, making its use alone rather precarious. However, Keating's patent[64] suggests that his time lock was intended for use on double-dial single-custody safes. With one time lock on each dial, the user had the reliability of two independent time locks.

Keating's time lock was probably intended as a small, low-cost competitor to Herring's Infallible time lock, good enough for use in fireproof and low-cost burglarproof safes, able to fit where others could not, but never intended to be an impressive piece of machinery. However, Yale was unlikely to have been interested in producing low-cost, low-margin time locks at a time when its high-cost, high-margin models were selling well. Once Herring's Infallible was forced off the market by lawsuits by Yale and Sargent, there was no need for a competitive model and development of Keating's time lock was discontinued. The example shown here is the only one known and may have been Keating's own working model. The design is not thought to have gone into production.

Keating's time lock. The smallest time lock ever made, Thomas Keating's time lock was made for Yale, probably to compete with Herring's Infallible, another small, round, case-mounted design.

Keating's time lock was designed specifically for use with a Linus Yale, Jr., single-dial combination lock. It is seen here mounted on the wheel pack, much in the same way as Herring's Infallible. As with Herring's Infallible, the Keating time lock would require that the safe mount two combination locks, or the risk of lockout would be too great.

SARGENT & GREENLEAF'S VARIANT MODEL 2 TIME LOCK

By the late 1870s time locks were sufficiently widely accepted to create demand for more customized formats. One example of such specific designs is a Variant Model 2 made by Sargent & Greenleaf around 1878. The movement is identified as that of a Model 2[6]: the movement serial number (no. 841), case serial number (no. 830), and the construction style (black dials without plate engraving or Geneva stops) place this time lock between Sargent's introduction of its Models 2[6] and 2[7]. However, the door hinge is of a larger, earlier style, similar to the Model 2[4]. The case is a one-piece casting, suggesting that it was factory-produced, possibly using an older, heavier hinge mold for greater strength, given the long, narrow door. The shortened drop lever extends through a slot in the case and is tapped for a set screw, creating a bottom-release mechanism to operate on a safe lock or an unusual boltwork. While the style is similar to that used to actuate an automatic bolt motor, automatics had probably not been introduced at the time this lock is thought to have been made. The Variant Model 2 was installed in a safe, and it is shown here with its original mounting plate.

Even the most successful time locks of the nineteenth century had production runs far smaller than those after 1900, since the number of banks in operation between 1870 and 1900 was never very great. Consequently, large makers were more willing to design particular variants of their base designs for use in specific safes. One example was Sargent & Greenleaf's Variant Model 2, circa 1878, which did away with its internal bolt altogether, instead using its drop lever as a bottom release. Though this Variant Model 2 shows wear consistent with having been mounted and used in a safe, the specific safe that this time lock was adapted for is not known.

SARGENT & GREENLEAF'S MODEL 4 TIME LOCK

In July of 1878, Sargent & Greenleaf introduced the Model 4, a two-movement time lock, smaller than its Model 3 and even easier to fit into smaller safes and existing vault doors without sacrificing security.[65] The movement format is slightly smaller, but most of the size reduction came from the Model 4's cello-bolt, even more compact than that of the Model 3. The drop lever is mounted behind the two solid-dial, single-plate forty-six-hour movements and the engraving pattern was limited to a simple stamped-leaf and line design on the drop lever and bolt. The Model 4 was the first Sargent time

ABOVE AND OVERLEAF: *In among the boltwork of a safe door, space was always at a premium, and even with Sargent & Greenleaf's Model 3, many safes simply did not have enough room to add a time lock. Sargent's Model 4 offered all the security of its larger brethren (and, hence, carried the same $400 retail price), but in a size that could be retrofit to a new class of smaller safes.*

MODEL 4

lock to feature Geneva stops (visible below the winding arbors), which would not become standard for Sargent until some years later. This was also the first Sargent model to feature a split front plate, a nascent modular design allowing movements to be independently disassembled. Absent for the first time is the lever lock on the door, replaced with a handcuff-key type lock. By 1888 this type of simple door lock was standard on all Sargent time locks.

The forty-six-hour Model 4 was made until sometime between 1880 and 1882, when it was updated to a forty-eight-hour movement. Sargent made 365 of the forty-six-hour Model 4,[66] of which fifteen are thought to remain today. The

196

ABOVE AND RIGHT: *As time locks evolved, the maximum time period rose, with the industry standard reaching seventy-two hours by the late 1880s. Sargent & Greenleaf's Models 2, 3, and 4 generally offered the same features at any given time, as shown by this seventy-two-hour Model 4 from that period, still mounted on the door plate from the safe along with a portion of the boltwork.*

example shown here has a machine-turned finish rather than surface jeweling, suggesting that it may have been factory-refurbished during its life.

By 1889 the Model 4 had seventy-two-hour movements with Sargent's text-free wagon-wheel dials. This seventy-two-hour version also featured larger Geneva stops to the sides of the winding arbors and an additional mounting screw per movement, visible on either side of the drop lever. The example above, right, is shown with its original engraved beveled door glass, mounted on the door plate with a portion of the original boltwork, and dates to about 1889. Although as many as eight hundred of this plain-dialed seventy-two-hour Model 4 were made, only about twenty-five remain today.

The dial changes made in 1896 to "Sargent & Greenleaf Company" and in 1918 to "Sargent & Greenleaf, Inc." on the Model 2 were also made to the Model 4. Toward the end of Model 4 production, Sargent returned to a one-piece front plate with 120-hour movements. All versions of the Model 4 cost $400, just as much as the larger but equally secure Models 2 and 3. Sargent continued making the seventy-two-hour Model 4 at least through 1950, with total production eventually exceeding five thousand, making it Sargent & Greenleaf's most successful design. However, like all small-format time locks, the Model 4 was most commonly mounted in smaller safes and often junked or scrapped with the safe. Consequently, the Model 4 is still rather rare, with about two hundred examples of all versions surviving.

Surviving Sargent & Greenleaf catalogues list a Model 5 time lock, available between 1901 and 1927. The only difference between the Model 4 and this advertised Model 5 was in the dimensions, and though the size of both models changed over time the Model 5 seems to have been consistently smaller, if only slightly so, than the Model 4.[67] No production records or examples of the Sargent & Greenleaf Model 5 are known to have survived.

EDWARD STEWART'S TIME LOCK

Patented in August of 1878 by Edward Stewart,[68] the Stewart time lock is based on two E. Howard movements thought to have at least fifty-six-hour power reserves. The most easily notable feature of the Stewart is its oval chain mechanism, which makes one complete revolution each twenty-four-hour period with tabs along the chain that allow the user to set an "open" period during which the safe door would be unlockable each day. The Stewart also has a calendar mechanism, the cog at the upper center, that allows the lock to skip the open period on any preset nonconsecutive days each week.

Available in both bolt-blocking and bottom-releasing formats, the Stewart time lock is engraved with the names Samuel Atlee and Jacob Blackburn, co-owners of the Stewart Time Lock Co. of Fort Madison, Iowa, with Edward Stewart.[69] Twenty of the Stewart time locks were made by E. Howard in April of 1882,[70] and four are known to survive, including two in private collections and the example shown here, which was donated to the Mossman Collection by J. & J. Taylor,[71] a safe maker of Toronto, Canada.

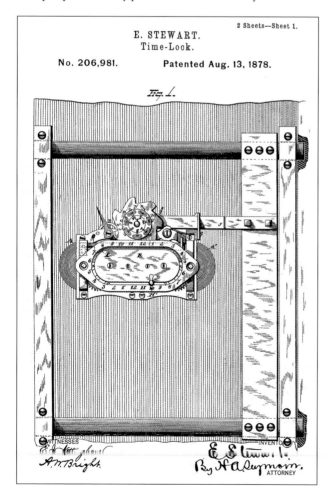

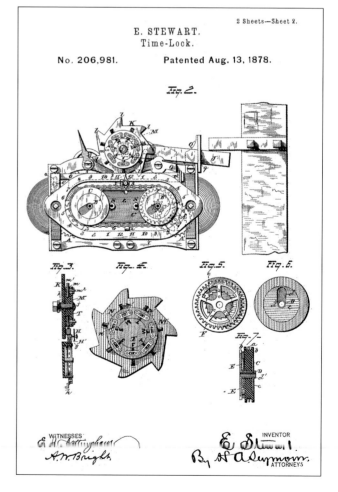

STEWART'S TIME LOCK

Stewart's time lock. The late 1870s was a period that saw new time lock designs patented and made, not only by smaller lock makers such as Lillie and North but by tiny, new companies with little more than a good idea. One such design was the Stewart time lock from the Stewart Time Lock Co. of Fort Madison, Iowa, with its tiny bicycle chain–like drive mechanism that let the Stewart lock and unlock automatically for a set period each day and skip its opening period on the days set on the gear at the top center.

OPPOSITE: *The patent drawings for the Stewart time lock accurately depict the mechanism in relation to the door's bolt, both locked (Fig. 1, opposite, left) and unlocked (Fig. 2, opposite, right).*

YALE LOCK MFG. CO.'S AUTOMATIC TIME LOCK

Around 1878, Yale Lock Mfg. Co. introduced what would be its last adaptation of Burge's Gothic time lock design, purchased in 1875, in an example of one of Stockwell's patents.[72] This extremely high-quality mechanism includes a small, interior bolt motor mounted behind the plate in the case top. A pair of springs attached to the boltwork through an opening in the case actively draws the boltwork into the case when the time movement releases, opening the safe door. Otherwise, Stockwell's patent operates on much the same principle as the Double Pin Dial, also a Stockwell design. Although the movements were made by E. Howard, they do not appear in the firm's surviving production records and this model is thought to have been discontinued prior to the start of Yale's time lock sales reporting ledger. Consequently, it is not known how many of this Stockwell time lock were made. This example is the only one known.

opposite: *Stockwell's patent, Yale's first automatic time lock. Mechanically related to Burge's 1872 Gothic time lock and Yale's 1875 wooden-case time lock, Stockwell's patent time lock also had aspects similar to Yale's Double (and later, Single) Pin Dial time locks.*

above: *Stockwell's patent included two springs to actively pull part of the boltwork. The connection is visible at the top right of the case with the spring-winding key in place. The winding arbors for the time movements are visible at the bottom left and right of the case and have their own, smaller key.*

SARGENT & GREENLEAF'S MODEL 3 TIME AND COMBINATION LOCK

Time locks were still a relatively recent option in the late 1870s but, with demand growing, many safe companies developed close working relationships with the best-known time lock suppliers. At this time, the Detroit Safe Co. was already a well-established safe manufacturer in the American Midwest and had pioneered the construction of a style of round-door pedestal safe, the forerunner of the "cannonball"-style safes later adopted by many safe companies and very popular among bankers in the 1890s. The Detroit Safe Co.'s close affiliation with Sargent & Greenleaf [73] was likely the basis for the Sargent & Greenleaf single-case Model 3, a smaller version of the Model 1, designed for a small, round safe door. With a bronze case, shield-shaped door, rollerbolt, and surface jeweling, the single-case Model 3 drew clearly on the Model 1 design. By substituting a two-movement, forty-six-hour, black-dial Sargent Model 3 to control a smaller Sargent combination lock, it was possible to fit this time lock into a small stand-alone safe door. While the single-case Model 3 may have been available for separate purchase, there is no record that it went into production. This is the only known example.

MODEL 3 TIME & COMBINATION LOCK

Sargent & Greenleaf's single-case Model 3. Similar in concept to its Model 1, this Sargent & Greenleaf time lock was smaller, using the movements from a Model 3. The round case was most likely designed to fit inside a round safe door made by the Detroit Safe Co.

BEARD & BRO. TYPE 1 TIME LOCK

Little is known about the St. Louis, Missouri, firm of Beard & Bro., a safe making firm of G. N. and E. J. Beard that produced its first time lock in 1878. Beard & Bro. employee Phinneas King secured his first time lock patent in February 1878[74] for a design based around an unusual gravity bolt that extended vertically through the top of the case. King's patent prototype was a full-featured lock with an industry-leading ninety-six-hour movement and a calendar mechanism. However, this model was never produced, probably due to prohibitive cost projections considering Beard & Bro.'s eventual advertisement of its time lock as the "Best and Cheapest on the Market," according to their letterhead of 1880.

The first time lock that Beard did produce drew in part on King's original patent but was based primarily on a second patent of March 1878.[75] This Beard Type 1 time lock uses two carriage clock movements with platform escapements visible through the two round case-top windows. At the time, even the finest carriage clocks used inexpensive brass for plates and platforms. Yet the movements in Beard's Type 1 used silver plates and platforms, an incongruous extravagance for a lock advertised on price. The case is plain nickel-plated bronze with rectangular door glass. It is not known how many of the Beard Type 1 were made, but it is thought to have been in production for only a short time before being replaced by Beard's Type 2. The only known surviving example of the Beard Type 1, shown here, was found in a shed in Michigan's Upper Peninsula in 1998 and has been restored to presentable condition.

Beard & Bro. Type 1 time lock. The first time lock made by Beard & Bro. was based on two patents by Phinneas King using two carriage clock movements with horizontally positioned escapements, only the sides of which are visible here. When the lock was in use, the escapements could be monitored through two round windows in the case top. The oblique armature pivots around the set screw visible at the bottom center, clearing or blocking the opening in the right side of the case.

BEARD & BRO. TYPE 2 TIME LOCK

In November of 1878, Beard & Bro. placed its first recorded order with E. Howard for twenty five time locks using its revised Type 2 design.[76] The Type 2 continued some elements of the Type 1, including the two forty-eight-hour movements, twenty-four-hour central dial, and the release mechanism, but included a number of significant changes, some seeming to conform to the production style of E. Howard. The Type 2 used two movements mounted on a single plate, rather than the two independent movements of the Type 1. With the new Howard movements' forward-facing escapements, Beard did away with the case-top glass in a more rounded nickel-plated bronze case, now with open crosshatch damascening. The front door glass took on the standard E. Howard–made "camel-back" shape found on Yale, Stewart, and, later, Holmes time locks. This same camel-back door glass shape was inverted on the Beard Type 2. Of the 175 Type 2 time locks Beard & Bro. ordered from E. Howard between 1878 and 1885, five are known to survive today. The example shown here from the Mossman Collection was adapted for use with an automatic bolt motor by the addition of the bottom release extending through the case bottom.

Between November 1885 and 1887, Beard & Bro. placed its last two orders with E. Howard totaling fourteen pieces, twelve time locks and two time lock models, of a revised third type. Substantially the same as the earliest Mosler design, only one of this Beard Type 3 survives today, in the Harry Miller Collection. The similarity to the Mosler stems from Beard & Bro.'s 1887 sale of its entire time lock business to Mosler, with Phinneas King apparently moving to Mosler as well.

BEARD & BRO. TYPE 2

Beard & Bro. Type 2 time lock. Like many other time lock makers, Beard & Bro. soon found that the best balance between cost and reliability could be found with time movements made by E. Howard. The company's Type 2 time lock did away with the peculiar carriage clock movements, replacing them with common forward-facing E. Howard movements that could be monitored through the door glass. The example shown here has a bottom release, using the side-release armature to add release power with a coil spring attached to the case bottom.

HOLMES TIME LOCK CO.'S HOLMES ELECTRIC TIME LOCK

In the late 1870s the Holmes Electric Time Lock Company debuted its first of four models, the Holmes Electric time lock. Charles Chinnok, a Holmes employee, first applied for the patent underlying the design in 1876, but it was not granted until 1879,[77] and though the Holmes model may have been made prior to the patent date no examples from the pre-patent period are known. Interestingly, engraving on the Holmes Electric also claims an 1872 patent[78] assigned to Holmes by Isaac and Abraham Hertzberg for an "apparatus automatically regulating flame of gas burners." The relevance of a patent for an automatically regulated gas burner is not entirely clear, although this may have been an attempt at patent priority. Throughout the time lock–related litigation, a number of diverse patents were relied on for priority with little success.[79] The first recorded movement production order was from E. Howard in July 1879, for 105 time locks.[80]

The Holmes Electric used a spring-loaded bolt block that released an opening in the upper right side of the case. The two E. Howard movements are of the same dimensions that E. Howard used for Yale's Double Pin Dial, and they are so close in design that many parts are interchangeable. A small "disabling latch" can be seen on the lower right of the front plate, which keeps the time lock from closing, avoiding accidental time-locking of a safe during the business day. Such a latch would not appear on a Yale design until after 1900 or on a Sargent design until after 1920. The two movements turn a single twenty-four-hour central dial that sets an opening period during each day, allowing the user to set the daily lock and unlock times once and simply wind the time lock thereafter.

Among the secondary time lock makers, Edwin Holmes was, for a short period, the most successful and his time locks were used by a number of high-profile clients, including the United States Treasury Department. Though its two redundant movements were more than sufficiently reliable to ensure unlocking, the Holmes Electric added a unique feature: a pair of electric solenoids visible at the top of the mechanism that could open the time lock should both movements fail. The example shown here still has the original bolt pivot plate attached to the right side of the case near the release.

The major innovation in the Holmes Electric was its backup system, a two-pendulum electromagnetic mechanism visible across the top of the case. During installation, a wire was connected to the contact visible on the top of the case and run to a contact on the exterior of the vault door. In the event that both timers failed, the external contact was connected to an independent pendulum that was manually moved back and forth, sending pulses to two solenoid coils in the time lock that advanced the movements until the lock released.

By 1881 Holmes had introduced his second version,[81] the New Electric time lock. This name would remain unchanged on all future Holmes models. The earliest type of the New Electric added a sixty-minute dial to allow the user to set the lock and unlock times more accurately. More than 125 of this Type 2 are thought to have been made between 1881 and 1882, with three known to remain today.

In 1882, Henry F. Newbury, a Holmes employee, found that Yale and Sargent time locks were vulnerable to a small charge of dynamite detonated against the safe door. Such an explosion could either release or disable these time locks. Soon after, Newbury was granted a series of nine anticoncussion patents.[82] By 1882 Holmes was producing his third model (the second style of New Electric or New Electric[2]),[83] which introduced a seventy-two-hour movement, a "weekend" mechanism that allowed the time lock to skip one or two days of scheduled opening, and a small button on the outside of the case door allowing the disabling latch to be operated without access to the time lock itself. More important, the

The only known photo of a Holmes Electric time lock installed in a vault door (bottom center), this illustration shows the solenoid backup system connected through the door by a single coiled wire. Conveniently, mounting room was not a limiting factor in this spacious vault, where the Holmes Electric was put on its own plate attached to the doorjamb by six tall spacer bolts that helped guard against objects striking the protruding time lock. The connection of the side release on the bolt pivot plate can also be seen. (Courtesy of Dean Cross.)

In the event that both Holmes Electric time lock movements stopped without releasing the bolt, this electric pendulum could be connected to the external contact. The pendulum would be manually swung, touching the contacts on either side, sending pulses to the solenoids in the lock and advancing the movements. (Courtesy of the Harry Miller Collection.)

New Electric[2] included Newbury's antidynamite technology, and attached to each Holmes New Electric[2] was a small decorative flange carrying two April 1882 patents and one January 1883 patent dates. None of these dates matches any patent on record with the Patent and Trademark Office, but they may refer to the anticoncussion patents. There is some speculation that Holmes intentionally misdated the plaque. While Newbury did have patent protection, this was useful only if Holmes could identify infringing time locks—a difficult task, since Newbury's antidynamite designs were subtle changes in the internal construction of the time lock. Hence, Holmes may have included these dates to advertise that the technology was patented while trying to keep the published details obscure. A fourth 1884 patent date on the flange is for the New Electric[2] design itself.

Holmes began making his fourth model, the New Electric[3], as early as 1884 but possibly in 1885. It had a smaller and more resilient electromagnetic mechanism. This New Electric[3] kept the standard case but used plain screws in all mounting holes rather than the knurled finger screws of earlier types. In addition, the locations of the electromagnetic pendula were changed from inside the case to outside the case set into depressions between the back of the case and the vault door. This curious change was possibly made to further counter the use of explosives. Holmes never advertised any of his time locks as antipowder, anticoncussion, or dynamite-proof, nor did anyone else. While no period documentation explains this, it may be due to an inherent suseptibility of delicate time pieces to explosive force that no maker was able to overcome with any degree of certainty.

During Holmes's production of the New Electric, it was such a highly regarded mechanism that it was used by both E. Howard[84] and the U.S. Treasury Department.[85] In total, Holmes made about 350 time locks. Today, ten are known to remain, of which five are the New Electric[3].

HOLMES ELECTRIC TIME LOCK

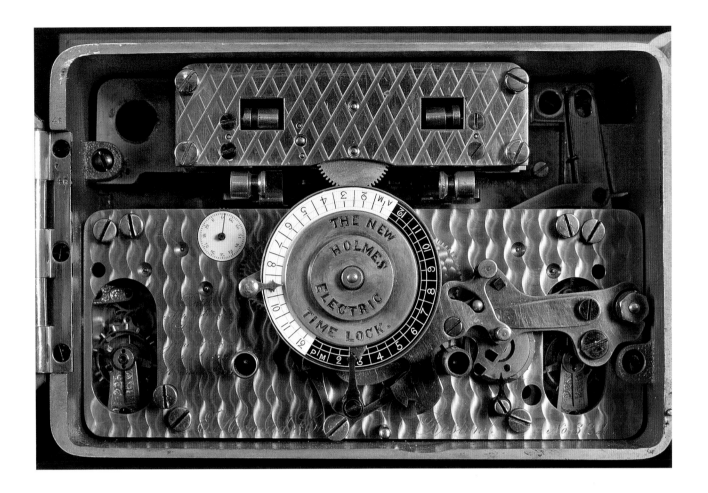

Holmes's last time lock, the Holmes New Electric[3], debuted in 1885 and included all the security and backup features of the three earlier Holmes designs, including the antidynamite improvements of Henry Newbury. These were often subtle spacing changes that buffered important parts from explosive shock. This time lock was so highly regarded that the E. Howard company—the makers of many time lock movements for major companies such as Yale and Hall's Safe & Lock—chose to use the Holmes New Electric[3]. Unfortunately for Holmes, this corporate snub led to a patent infringement lawsuit by Yale and the end of the Holmes Electric Time Lock Company.

MILTON DALTON'S TWO-MOVEMENT TIME LOCK

Hall's Safe & Lock Co. invested heavily in research and development of its time lock designs. Yet despite having a number of successful time lock innovations by 1880, Hall's general manager of business[86] and ingenious inventor Milton Dalton continued his unending search for time lock improvements. One rare design of Dalton's that was apparently never patented is seen here. Based on two forty-eight-hour E. Howard movements, this prototype had a bottom release to work with a combination lock, a jeweled bronze case, and a nickel-plated interior plate with open crosshatch damascening.[87] A second, incomplete example is also known to survive with the early Burton-Harris automatic bolt motor that it controlled.

Milton Dalton was a prolific and important inventor at Hall's Safe & Lock Co., with many important patents during his lifetime, but not all Dalton designs saw success. This Dalton time lock's security and reliability were probably the match of any contemporary, yet it does not seem to have been patented or produced.

CONSOLIDATED TIME LOCK CO.'S TWO-MOVEMENT TIME LOCK

Prior to 1882, time locks made by Joseph Hall's employees were attributed to "Joseph Hall" or "Hall's Safe & Lock Co." Then, in January of 1880, Hall incorporated the Consolidated Time Lock Co., an independent company owned by Joseph Hall, to both be Hall's time lock supplier and insulate Hall's stable and successful safe and lock business from the commercial and legal risks of the new time lock market.

The first time lock produced under the name Consolidated appeared in 1882, a two-movement Consolidated made only from January through October of 1882,[88] identifiable by a nickel-plated bronze case about half an inch wider than later models, with a floral engraving pattern on the door and top. The two forty-eight-hour movements in these early time locks have a concentrically circular engine turning on the front plates, which was later replaced by the more common wavy vertical damascening. One hundred of this model were made[89] but only two of the earliest wide-cased Consolidated time locks are known to survive today, both with bottom-releasing mechanisms for use with a Hall Premier safe lock.

The example shown here is elaborately engraved with a star-and-stripe-patterned shield on one side and a view of a sail boat from a veranda on the other. This notable style of folk art engraving on the cases is found sporadically on Con-

BELOW, OPPOSITE, AND PAGES 216–217: *The Consolidated Time Lock Co. was founded in 1880 by Joseph Hall to insulate his successful safe and lock business from the legal and market risks of the new time lock market. The first Consolidated model was a two-movement model with the bottom release common to the vast majority of Joseph Hall's time locks. Though the movements are horizontal, they are not the carriage clock movements found in the earliest Beard & Bro. model. Rather, they were standard E. Howard movements adapted for this low-profile design.*

OPPOSITE, BOTTOM: *The case top of the first Consolidated model featured round beveled glass to monitor the horizontal movements. Some also had the elaborate vine engraving patterns seen here that were part of a series of folk art-style engravings on the case sides.*

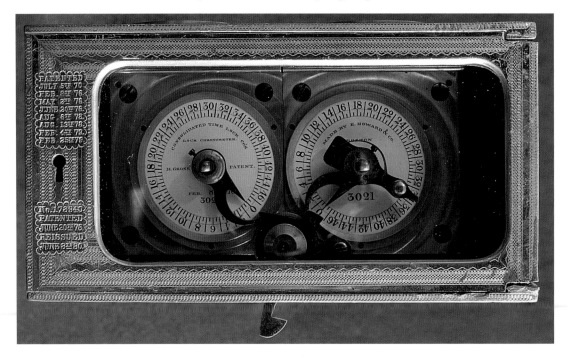

solidated and Dalton (also part of Hall's Safe & Lock) cases from this period, continuing through about 1886. The theme to the engravings may be the Ohio River, with subjects known to include a boat towing another boat, ducks, flags, cranes, boys fishing, and a country gentleman with a fishing rod. The work may have been done by the same artist, given the consistent appearance of the associated floral vine engraving. Eight folk art–engraved examples are known to survive. Later case-side engraving was machine-done in a circular guilloche pattern.

The vine engraving patterns are also found on single-movement Consolidated time locks that arrived later. These folk art–engraved time locks seem to have appeared intermittently between 1882 and 1886.

The subjects of the folk art engravings were varied, depicting a number of scenes likely common in the American Midwest of the late 1800s. Aside from the shield, the images seem to center around the Ohio River, which runs through Cincinnati, depicting boats, people fishing, and birds common to wetlands.

CONSOLIDATED TIME LOCK CO.'S BENJAMIN FLINT TIME LOCK

In 1883, the Consolidated Time Lock Company produced a time lock patented by Benjamin Flint[90] that dogs the boltwork rather than the attached combination lock. This was an unusual design for Consolidated, a company with more than ten years of success marketing time locks that controlled an attached combination lock. The design features the single-movement Infallible lockout protection system. The mechanism to act directly on the boltwork[91] required the addition of the armature projecting from the lower right corner of the time lock case.

This example is paired with a Joseph Hall Premier five-tumbler safe lock and was mounted on an interior safe door. There is no known record of how many of these double-acting time locks were made by Consolidated. This is the only known surviving example.

BENJAMINN FLINT TIME LOCK

The Consolidated Infallible time lock with the Hall Premier. The Consolidated Time Lock Co.'s Infallible mechanism was so reliable and effective that it became the basis for the vast majority of the company's models. Even when Consolidated produced an unusual time lock that operated on the boltwork rather than on the combination lock, the Infallible mechanism was part of it, using both a side release and a connection to the combination lock rather than a second movement.

ABOVE: *With the cover of the Premier combination lock removed, the lower portion of the Infallible linkage is visible. Though the parts were nickel plated in some examples, in this one they are blued and bronze, extending down from the time lock and around behind the wheel pack mounting bracket.*

GEORGE DAMON'S BANK LOCK

George Damon and his Damon's Bank Lock Co. were well known and well regarded in the bank lock industry in and around Boston, Massachusetts, from before 1870 until after 1895. During this long period of successful operation, Damon introduced a number of combination lock designs that were used often on bank vaults of the time. One of Damon's attractive models was this key-changing four-tumbler lock that, despite its significant size, was usually mounted in pairs. Fewer than twenty of these large Damon combination locks are known to remain today. This example was donated to the Mossman Collection by Edwin Holmes.

GEORGE DAMON'S BANK LOCK

This combination lock from Damon's Bank Lock Co. first appeared in the mid-1880s. Its key-changing design was made for the largest safes and vaults, requiring greater than average space to accommodate its large size when mounted in pairs, as intended by Damon.

YALE & TOWNE SINGLE PIN DIAL TIME LOCK

Between 1875 and 1883, the only small-format time locks available were the Sargent & Greenleaf Model 4 and various models from Joseph Hall and Consolidated Time Lock. During this period, Yale Lock Mfg. Co. made only the Double Pin Dial time lock, an excellent and full-featured design but certainly not small format. In 1884, Yale & Towne Mfg. Co. (renamed after the 1883 death of Linus Yale, Jr.) began producing a Single Pin Dial, a reengineered version of the Double Pin Dial that had all the same functions but was based around a single central dial, affording significant space savings.

The Single Pin Dial was made with two E. Howard movements capable of seventy-two-hour operation, the power reserve that was becoming the industry standard. It also featured a calendar mechanism that displayed the day of the week on a central white enamel bezel that was initially affixed with two screws, but with four screws after 1890. The case and movement's front plate are nickel-plated damascened bronze and the front plate is split to allow disassembly of one movement at a time. A brass gear on either side of the pin dial and a white dial above show the number of hours and minutes remaining until opening. Like the Double Pin Dial, the arbors are wound through eyelets in the door glass.

As with the small-format Sargent models, Yale's Single Pin Dial sold for $400 ($450 with a Sunday Attachment) the same prices as its Double Pin Dial, but this model never enjoyed the widespread popularity of its Sargent counterparts. Fewer than five hundred were made between its 1884 introduction and 1900, when production ended,[92] with most sold to small, midwestern banks that operated with smaller safes and vaults.[93] Today, fewer than twenty-five examples of Yale's Single Pin Dial time lock are known to survive.

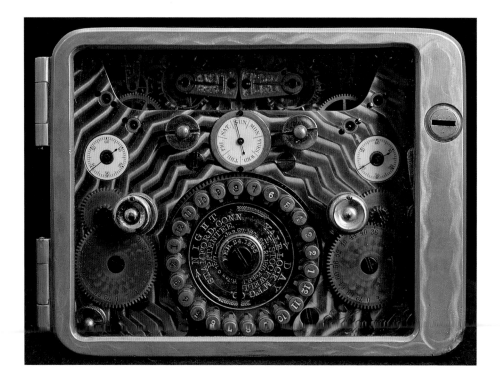

YALE SINGLE PIN DIAL

In 1884 the Yale & Towne Mfg. Co. introduced the Single Pin Dial, hoping for popularity similar to that of Sargent & Greenleaf's smaller-format Models 3 and 4. Offering all the features of the Double Pin Dial in a more compact design, the Single Pin Dial was one of the most intricate and well-engineered time locks to see significant production. However, for reasons that are unclear, the Single Pin Dial was never broadly adopted and fewer than five hundred would be made during its sixteen-year production life.

CHICAGO SAFE & LOCK CO.'S GEM TIME LOCK

On March 19, 1885, E. Howard recorded its first order of time locks from the Chicago Safe & Lock Co. of 209–217 South Canal Street, Chicago, to be numbered from 501 to 600.[94] Although it was already rather late in the industry's development for a new time lock company to hope for major success, Chicago Safe & Lock had secured the rights to produce a new design from established inventor Henry Gross, who had recently left Hall's Safe & Lock Co. Gross had been an important inventor for Joseph Hall's company and his patents formed the basis for Hall's Safe & Lock company's success. However, a disagreement had developed between Gross and Hall over patent royalty payments that eventually led not only to Gross's incensed departure but also to his decision to testify on behalf of Sargent and Yale in subsequent patent infringement litigation.[95] As was common industry practice, Chicago Safe & Lock placed its production order with Gross's patent application still pending, and when it was awarded on April 14, 1885,[96] Gross assigned the patent to Chicago Safe & Lock. On December 11, 1885, Chicago Safe & Lock placed an order with E. Howard for twenty new dials for these time locks that bore the marking Gem Time Lock Co., Gross Patents, Chicago, Ills.[97] Thanks to this change, possibly an attempt to avoid lawsuits, this lock is known as the Gem time lock.

The two-movement Gem bears more than a passing resemblance to Gross's two earlier time locks. The front plate, back plate, and mainspring are nearly identical, and although the gears and platform escapements are the same size as those in Hall time locks, they are not interchangeable. New to this design was the single arbor that winds both movements simultaneously, a feature that would appear again in the next generation of Yale time locks beginning in 1887. The Gem was one of the first time locks intended to be used alone, with an automatic bolt motor rather than a combination lock.[98] There is some indication, however, that the Gem required a specially installed boltwork,[99] which would have had a detrimental effect on sales, as with Sargent & Greenleaf's Model 1.

Interestingly, E. Howard's production records reveal that Chicago Safe & Lock ordered only seventy cases with their order of one hundred movements. The earliest Gem time locks are thought to have been assembled in unmarked cases discarded by Hall Safe & Lock. Chicago Safe & Lock placed another order for one hundred Gem time locks numbered from 601 to 700 on May 31, 1888.[100] Of the two hundred made, six surviving examples of the Gem time lock are known today.

CHICAGO SAFE & LOCK CO.'S GEM TIME LOCK

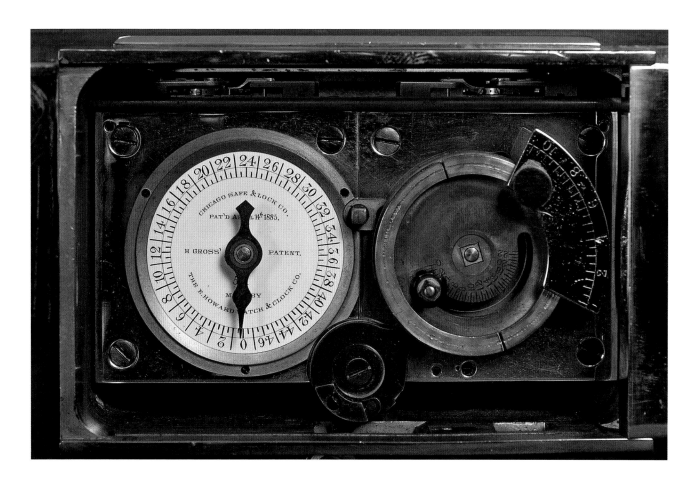

Gross's Gem time lock. Henry Gross began his career with Hall's Safe & Lock Co. but eventually left and joined the Chicago Safe & Lock Co. for a short period, during which he introduced the Gem time lock in 1885. The interior view shows the single arbor that winds both movements.

CHICAGO TIME LOCK CO.'S PERFECTION TIME LOCK

Following the success of the Gem time lock, Henry Gross, E. W. Neff, and a group of investors founded the Chicago Time Lock Co. to produce and sell Gross's newest design. On March 1, 1886, Chicago Time Lock placed an order with E. Howard for one hundred "double Neff movements" to be made with gold hairsprings and serial numbered from 1.[101] The result, based on Gross's latest patent,[102] was the Perfection time lock, one of the smallest high-security time locks ever.

This first type of Perfection, serial numbered 1 to 100, had a nickel-plated, fully removable bronze lid and two forty-eight-hour movements housed in a round, nickel-plated bronze case. The dials are numbered through forty-six hours, but the indicator can be wound fully back to "0" for a total running time of forty-eight hours. The trioval-shaped metal escapement covers rotate around hinge pins, revealing the escapements for observation. The winding arbors abut the enamel dials and a single mounting hole runs through the case center. Three examples of this first type of Perfection time lock survive today.

At some point after 1890, Chicago Time Lock designed a revised version of its Perfection time lock, retaining the major aspects of the design but incorporating a number of minor improvements. The lid was now hinged and the metal escapement covers were replaced with glass bezels. The winding arbors were positioned further from the enamel dials, possibly to prevent damage to the dials, and the central mounting hole was replaced with a mounting plate. The Perfection still featured two forty-eight-hour movements, which were now made by an unknown maker rather than by E. Howard. Of the approximately fifty of this second style of Perfection thought to have been made, the only complete example remaining is shown here. Five other partial examples are also known.

ABOVE AND RIGHT: *The Perfection time lock. This was one of the smallest and best-made time locks of the day, with a compact design based around a pair of movements specially made by E. Howard. These "double Neff movements" were made in pairs, sharing a serial number. The mechanism was the picture of simplicity, with a single central lever that blocked the cutout opening at the top of the round case, although the movements could not be monitored without both removing the cover and swinging back the trioval lids.*

PERFECTION TIME LOCK

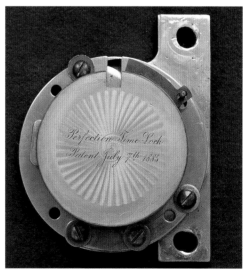

ABOVE AND RIGHT: *The Chicago Time Lock Co. produced a revised version of the Perfection time lock, probably after 1890. The basic design remained with only minor functional changes such as a hinged lid, fixed-glass movement lids, and winding arbors placed farther from the delicate enamel dials. While its simplicity of function remained, the Perfection time lock also retained its forty-eight-hour duration, a period shorter than the now industry-standard seventy-two hours, which may have limited the Perfection's popularity.*

CONSOLIDATED TIME LOCK CO.'S E. J. WOOLLEY TIME LOCK

In 1886, Edward J. Woolley was awarded a patent that became the basis for a two-movement time lock later that year. Woolley's patent time lock was made by the Consolidated Time Lock Co. based on its popular and widely accepted forty-eight-hour E. Howard two-movement design. Unlike other designs by Hall or Consolidated, however, Woolley's design operates directly on the boltwork rather than on the combination lock. This allowed for it to be used alone, without a key or combination lock, which had become a popular format for safes used daily in commercial operations. With no combination lock, there was no way to use Hall's Infallible device; hence, this model was always made with two movements and featured factory-original bolting flanges for precise positioning to ensure proper operation. Fewer than one hundred are thought to have been made between 1886 and 1887. A number exist in part but the example shown here is one of only four complete examples known to survive.

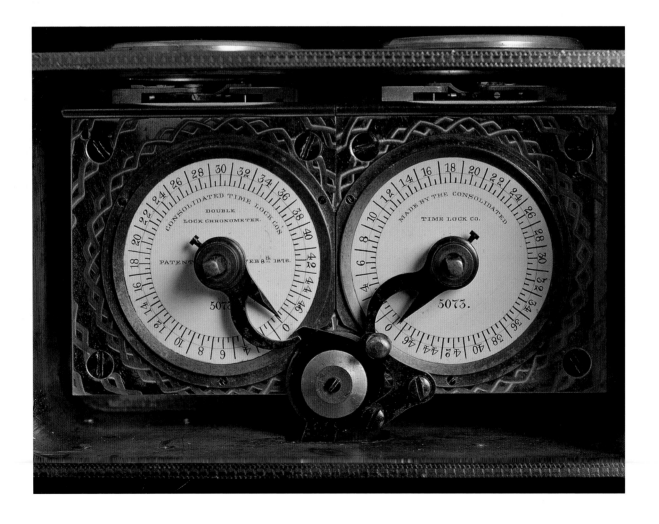

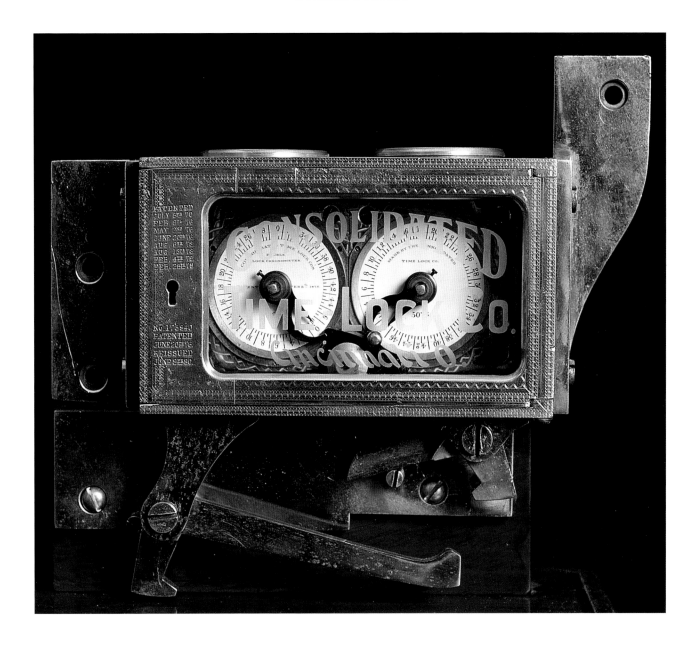

Regardless of the Consolidated Time Lock Co.'s penchant for using a time lock in conjunction with a combination lock, there was always demand for independently operating time locks, and in 1886 E. J. Woolley designed his patent time lock for Consolidated. Woolley's patent time lock included two E. Howard movements like the earliest Consolidated models, but in place of the simple bottom release it used a bolt-actuating mechanism below the case to release the boltwork. Without a bushing in the case to control this piece, proper positioning on the safe door was important, leading to mounting flanges being built into the case.

SEDGEWICK'S ELECTRIC COMBINATION LOCK

Even though the combination lock had firmly taken hold in the American bank lock industry by the mid-1880s, with key locks relegated to private and small-scale commercial uses, safe and combination lock makers were always aware of the one inherent weakness of a dial combination lock: its spindle. While major strides had been made ameliorating this weakness (e.g., offset, indirect, and flanged spindles) ultimately there was a risk inextricably associated with having a dial outside the safe physically connected to the lock inside.

With the arrival of urban electrification in the 1880s, it was only a matter of time before a lock maker sought to substitute the spindle with an electric control. In 1886 Sedgewick, an employee of Herring & Co., was awarded a series of four patents for his Electric combination lock.[103] Later Sedgewick designs incorporated compressed air mechanisms, but the only complete example known to be made used electricity only. Shown here, Sedgewick's Electric kept the now common dial, but with no physical connection to the lock's interior Sargent's micrometer was entirely useless, and there was no clue as to the location of the lock for drilling.

Sedgewick's Electric is not known to ever have been sold or used, and there is no firm evidence that any electric lock was offered for sale before Yale's 1896 Electro-Magnetic combination lock.[104] Yale's Electro-Magnetic used a design different from Sedgewick's Electric and is not known to ever have been installed. During this era, electricity was a new science and, like the time lock, a major change would be needed to convince a conservative nineteenth-century banking industry to adopt such a fly-by-wire design. It was not until the 1970s, when the fine machining needed for top quality locks had become so expensive and electrical circuitry so cheap, that electrically controlled bank locks became a major design competitor for the first time. The Sedgewick Electric shown here is the only known surviving example, made by Herring and donated to the Mossman Collection by the Herring-Hall-Marvin Safe Co.

LEFT, ABOVE, AND OPPOSITE: *Sedgewick's Electric safe lock. The first electrically actuated combination lock was probably this 1886 design patented for Herring & Co. Five wires ran from the dial to the lock, one for each opposing pair of oval electromagnets and a grounding wire. And though Sedgewick's Electric did away with the weakness of a spindle, many people were cautious about adopting far more mundane electric devices. Sedgewick's Electric is not known to ever have been produced, sold, or used.*

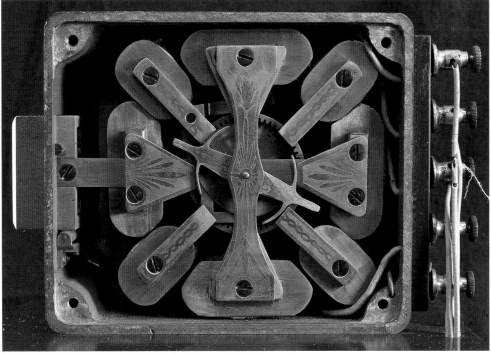

MOSLER SAFE & LOCK'S 1887 TIME LOCK

The time lock shown here is listed in *Lure of the Lock* as a Mosler Calendar time lock,[105] an understandable attribution, given the Calendar time lock's place at the forefront of Mosler's renown when *Lure* was published in 1928. However, this lock predates Mosler's Calendar by about four years and does not have the calendar feature. Rather, this lock was made in 1887, delivered from E. Howard to Mosler Safe & Lock Co. as part of an order for one hundred at $78 each placed in April 1887.[106]

That order was the first for Mosler Safe & Lock, which had bought all rights to Beard & Bro.'s time lock design earlier that year. Mosler left the design substantially unchanged from Beard's last variation, using two forty-eight-hour E. Howard movements controlled by the twenty-four-hour central dial. Small changes include through-door winding eyelets, a different door keyhole, no damascening on the central dial, and a nickel-plated case interior rather than the industry classic scarlet paint. This early Mosler time lock was rather successful with more than four hundred made based on King's patent.[107] Most were sold as part of a banker's money chest featuring Mosler's patented screw door. It was available with either a bottom release for an automatic or a side release for the boltwork, and toward the end of its production the gravity-driven side release was augmented by a coil-sprung piston mounted on the bottom left corner of the front plate, maintaining downward tension on the controlling lever. Of the four hundred of this earliest Mosler Safe & Lock time lock made, only five are known to have survived.

The Mosler Safe & Lock Co. sought to enter the time lock industry relatively late by buying the time lock portion of Beard & Bro. in 1887. Mosler's first model was substantially unchanged from Beard's last. One difference found later was the spring piston at the lower left that replaced the coil spring that the Beard model placed on the right, between the end of the armature and the case bottom. Present in the original patent by Phinneas King but absent here is the calendar mechanism, a seven-toothed gear that would have been located over the two-toothed gear at the lower right.

CONSOLIDATED TIME LOCK CO.'S
DALTON TRIPLE GUARD TIME & COMBINATION LOCK

In 1888, Consolidated Time Lock introduced a single-case format time and combination lock, the Dalton Consolidated Triple Guard. A seemingly after-the-fact competitor to the early single-case designs of Sargent and Yale (available from approximately 1876 to 1885), the Triple Guard is unquestionably the most complicated vault lock ever made. It featured a unique Milton Dalton–patented[108] single-tumbler combination lock that could be opened with a combination of anywhere from one to six numbers, set with the lateral lever across the single tumbler disc, controlled by a Dual Guard time lock.

The Dual Guard was first produced in 1884[109] and was independently successful, being sold on its own for use with combination locks made by MacNeale & Urban as well as by Hall. The Dual Guard used an unusually expensive E. Howard movement and, at $55 wholesale, its single-movement cost exceeded that of the two-movement Yale Double Pin Dial. The Dual Guard was equipped with its own internal combination lock, allowing the Infallible mechanism to be opened with from one to four numbers by shifting the pin visible below the time lock dial. Since the Dual Guard was intended for use with locks other than Hall's Premier, it used a gear on the case back to connect its miniature combination lock to the main combination lock rather than the armature found on Hall's Premier. Two styles of the Dalton Dual Guard were made: one that engaged the bolt with a plain hinged armature and a second that used the wound spring mechanism seen here on the outside of the time lock case. (Notably, even inside the Triple Guard, the case of the time lock shows hinge holes for its own door.) Of the independent Dual Guard time lock, fewer than five hundred are thought to have been made and fewer than ten are known to survive.

DALTON TRIPLE GUARD

The Dalton Consolidated Triple Guard was possibly the most complex bank lock ever, certainly for its time. The innovative combination lock could be set to require anywhere from one to six combination numbers alternating with a fixed reference number. This lock has only a single turning disc with small pegs in its holes corresponding to the combination numbers. As the disc is turned back and forth, the armature is lifted upward by discrete amounts, eventually releasing the bolt that runs below the disc.

Like the other single-case designs, the Dalton Consolidated Triple Guard saw limited use, at least in part due to its cost, estimated at between $500 and $600. However, unlike the Sargent and Yale single-case designs, the extreme complexity of the Triple Guard's construction and operation are likely to have been additional impediments to its adoption. Although the Triple Guard was described in Dalton Consolidated catalogues, it had long been believed that the model was so prohibitively complicated and expensive it was never made. However, an example was found in a private collection in 2001, serial numbered 2, that shows wear consistent with having been mounted and used in a vault.

The time lock portion of the Triple Guard was also sold individually as the Dual Guard, a single-movement time lock that featured the Infallible backup combination and was available with an additional spring assist seen outside the case on the right. The Dual Guard's Infallible mechanism had its own specially designed geared connection to the spindle. This modified Infallible operated a minute combination lock wholly included in the time lock. The tiny lock is visible behind and to the right of the time lock's dial.

CHAPTER 5

The Era of Monumental Security: 1888–1899

As the nineteenth century drew to a close, banking had grown from local to big business and banks began to look for ways to present an image of security and stability. One well-known example was J. P. Morgan's Wall Street office building. Located on one of the most expensive pieces of real estate in the United States, it is only two stories tall—Morgan's statement of financial strength. Most banks, however, chose a monumental vault as a symbol of their strength, taking full advantage of the industrial revolution in the United States. Along with these massive safes and vaults came the demand for equally impressive locks. The combination and time locks of this period were notable for their strength and reliability, allowing bankers to secure with confidence some of the most imposing safes and vaults ever made.

SARGENT & GREENLEAF'S CORLISS MODEL 4

The Corliss Safe Co. was founded in 1878 by William Corliss, an industrialist with a background not in lock and safe making but in steam engine construction. His grand vision and financial wherewithal made Corliss Safe an instant success. The Corliss signature design had no recognizable door, but rather a concentric rotating core and shell. Corliss's signature model was the Planet Safe, an enormous safety deposit safe made in a forty-eight-inch, thirteen-thousand-pound model and a sixty-four-inch thirty-thousand-pound model.[1] The company's Spherical—the "smaller" model—was also a behemoth, available in thirty-two- and thirty-six-inch versions, weighing eight thousand and twelve thousand pounds, respectively.[2] It is not known whether the Corliss Spherical offered a time lock option prior to 1888, but in December of that year[3] Sargent & Greenleaf introduced the Model 4 single-movement time lock designed specifically for use with the Corliss Spherical.

Sargent's single-movement Corliss Model 4 was produced and sold in matched pairs. The examples shown here, serial numbered 1348-1 and 1348-2, were shipped from Sargent for use in a Corliss Spherical in January of 1899.[4] The jeweled bronze case housed a seventy-two-hour movement with a Geneva stop under the winding arbor mounted on an unengraved front plate with a plain escapement bridge. The nickel-plated drop lever, case-back plate, and nickel- and bronze-finished bolt are engraved and Sargent's patent dates are shown on the bolt. One Model 4 would be mounted on either side of the Spherical's revolving inner core.

Serial numbering on Model 4 cases was consecutive, but the Model 4 was not limited to delivery to Corliss. Single-movement Corliss-specific batches were interspersed with double-movement Model 4s intended for individual sale to other safe makers. The first recorded set of Corliss Model 4 time locks was recorded as "1000 R+L" (i.e., serial numbered

CORLISS MODEL 4

A Corliss safe used a pair of Sargent & Greenleaf Model 4 time locks, each mounted in its own space cast into either side of the core opening. Although most Model 4 time locks were intended to work alone, specially numbered pairs were made for Corliss, each with a single movement.

1000-1 and 1000-2) through "1022 R+L." The next ran from 1056 R+L through 1068 R+L, with individual two-movement Model 4s numbered from 1023 to 1055.[5] Further, not all Corliss Model 4s had single movements. A Corliss publication included an image of a Corliss Spherical showing a two-movement Model 4,[6] but no examples of "R+L"–designated two-movement Corliss pairs have been found.

CORLISS MODEL 4

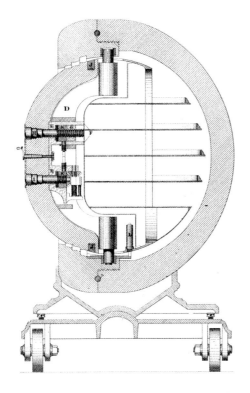

The large Corliss Planet was sixty-four inches wide and weighed fifteen tons. It is shown here with its opening handle in place. Once the locks were released, this handle would be turned, retracting the door into the body of the safe where it could be rotated to reveal the contents.

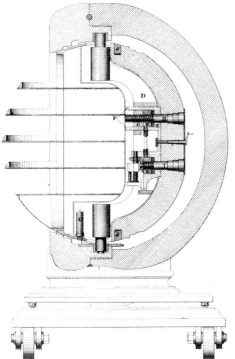

LEFT: *Cross-sectional schematics of the Corliss safe, closed and open. Here we can see why the Corliss design was so secure. With no real door, even recently developed high explosives such as nitroglycerin were of little concern since the entire circumference of the safe opening served to hold the core in place.*

OPPOSITE: *William Corliss, with his background in heavy industry, was one of the most innovative safe makers of the late nineteenth century. With its spherical shell and rotating core, a Corliss safe offered no corners, edges, rivets, or welds for criminals to exploit. The smaller Corliss Spherical is seen here partially opened. The core has been rotated slightly to the right, revealing one of the time locks embedded in its side.*

YALE & TOWNE TYPES B, C, D, AND E TIME LOCKS

In 1887, Yale patented a time lock design that would mark a new direction for Yale and, eventually, the entire time lock industry: its Type B and Type C time locks, which went into production in 1888.[7] Based on pocket watch movements rather than on the larger clock movements of the Pin Dials, these smaller-format movements were inherently suited to be individually replaceable or "modular" movements.

Modular movements were not original to Yale, having been a feature of Holbrook's Automatic time lock of 1858. However, Yale's reintroduction of the modular design would be enthusiastically copied by makers and lauded by users, since it significantly reduced the time and cost of maintenance and repairs. Maintenance or repair of built-in movements required a trained technician on site, whereas a modular movement could be replaced quickly by a layman with a clean, working movement and the problematic movement sent on to a central repair facility. This practice reduced costs but also resulted in time locks with nonconsecutive movement numbers and greater temporal differences between their movement and case serial numbers.

There does not seem to have been a Yale time lock designated "Type A." We find no photos or engravings, no mention in Yale period catalogues, no mention in any sales or production records, and no known examples of what could be a Yale Type A. The first of these new Yales seem to be the Types B and C, which differed only by their release, the Type B featuring a bolt dog and the Type C having a bottom release for an automatic. Both Type B and Type C used three watch movements, with the three independent power springs wound simultaneously by turning the surrounding seventy-two-hour rim counterclockwise. While simple and convenient, there was no chance to correct for overwinding. (With individually wound movements, setting one for too long a period was only a minor problem. The user could set the other(s) for the correct, shorter period and the lock would release whenever the first movement ran down.) Consequently, Yale's instructions noted specifically that the user must "be careful when winding to turn the cylinder to the left to take up the recoil of the springs, and to leave the desired mark standing exactly opposite the pointer."[8] But with the risk of the inconvenience of overwinding so high, Yale included an overwinding correction pin hole at the six- and nineteen-hour marks. Should the user overwind, a special pin was placed in the six-hour hole. The dial would then be turned to the number of intended locked hours plus six, and as long as the pin was in place the lock would open when the movements ran down to 6 rather than 0. The nineteen-hour hole worked in a similar fashion.[9]

Production of Types B and C was likely under way at least by the 1887 patent date, with sales starting later. Yale's records show that eighteen Type Bs were sold between December 1888 and June 1890 and fourteen Type Cs between May 1888 and

TYPES B, C, D, AND E

OPPOSITE AND ABOVE: *Yale American Waltham time lock. Yale's Type D and E time locks differed only in their release, the Type D having a side release at the top right (opposite) and the Type E a bottom release (above). Common to the Yale Types B, C, D, E, and later G time locks was a simultaneous winding mechanism. Rather than winding each movement individually, a single central arbor winds all three movements at once, ensuring an equal period on all of them. With this convenience came the drawback that, should the lock be overwound, all movements would be overwound. (With independently wound movements, the first movement to wind down would release the door, allowing the user to correct for an overwound movement.)*

With the cover removed from this Yale Type E, the elaborate damascening on the American Waltham Watch Co. Hillside movements and the seventy-two-hour edge markings can be seen more clearly.

March 1889. Serial numbering for the Type B seems to go up to 44, and to 40 for the Type C, but these numbers may not have been consecutive.[10] Yale also made and sold a Type BB, thought to have been a pair of Type Bs acting in concert. Two Type BBs were delivered to MacNeale & Urban in 1888, serial numbered 6 and 7, but there is no record as to whether numbers 1 through 5 were produced or sold.[11] There is no known mention of a Type CC. The rarity of Types B and C is not only due to their limited production but also because they were actively replaced by Yale after the introduction of its Types D and E.[12] Only one Type B is thought to be extant today; no Type C is known to survive.

Yale's Type D and Type E time locks were introduced by May 1889[13] and were intended to replace the Types B and C, respectively. Again, the only difference between Types D and E was the release, the D having a bolt dog and the E a bottom release.[14] While the basic features of the Types B and C were retained, Yale introduced substantive improvements in the Types D and E. Yale kept the simultaneous winding system but, notably, brought back a key-driven winding arbor. Now with a more accurate central arbor winding the three seventy-two-hour American Waltham Watch Co. pocket watch movements, the six- and nineteen-hour overwinding pin holes disappeared. A "stop pin" was introduced that could be pushed in to block the time lock from activating, should the user want to close the vault door with just the combination lock, a feature first introduced on the Holmes Electric. Yale also added a power reserve indicator for each movement, marked "up" and "down," a feature not available since the decline of the Pillard time lock in 1877.

Around 1890,[15] production of the Types D and E changed slightly. Prior to this, movements had a bimetallic balance wheel made of brass and steel, gold-plated up/down indicators, and no attribution to Yale anywhere on the lock. After about 1890 the balance wheels were plain brass, the steel up/

down indicators had no plating, and a Yale attribution appears on the rim of the cover.

Yale sold 62 of the Type D and 139 of the Type E between May of 1889 and June of 1892. Today, there are known to be fewer than five of the Type D and fewer than ten of the Type E. As with the Type BB, Yale offered a Type DD and Type EE, combining two Ds or Es in concert. While none of these is known to survive, records show that Yale did sell three Type DDs in March 1890 and thirteen Type EEs between November 1889 and December 1891. The Type EE is now thought to be what was described as the "Sextuple Time-lock" in Yale literature.[16] The Type E shown here, serial numbered 72, was installed in the People's Savings Bank of Grand Rapids, Michigan, on April 24, 1891.[17]

With the introduction of the Type D and Type E time locks, Yale debuted its own line of automatic bolt motors. The Type E shown here was paired with the most sophisticated of these early Yale automatics, with two fully redundant mechanisms. The Yale Type D and Yale automatics were soon popular with banks and businesses with regular hours, used in the new "solid door" safes, a safe with no key or combination lock that relied solely on a high-quality time lock and automatic bolt motor.

ABOVE AND OPPOSITE: *Unlike early bottom-release time locks that were usually paired with a combination lock, the bottom-release time locks of the late 1880s began to work with the newly developed automatic bolt motor. Often requiring a stronger release, these "automatics" housed powerful springs capable of operating the heaviest boltworks in the largest vaults. These gave rise to "solid door" models of safes and vaults—doors that had no keyhole, combination spindle, handle, or any other connection to the outside. The Yale Type D is shown here connected to a Yale No. 1 Double Reserve automatic that featured two redundant mechanisms, each wound independently with a long handle that attached to the seven-hole arbors at the left.*

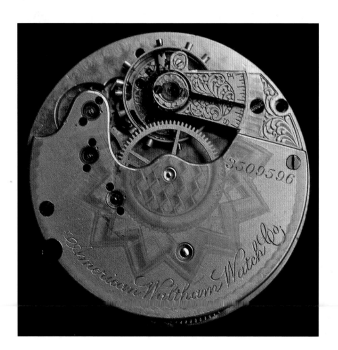

LEFT: *Yale's Types D and E time locks departed from the E. Howard–made clock movements in favor of the compact and easily changeable pocket watch movements made by the American Waltham Watch Co. seen here. While not yet truly a modular design, since replacement required skilled adjustment, the Waltham movements were a major step in this direction for Yale, whose previous designs had relied on built-in clock movements.*

TYPES B, C, D, AND E

YALE & TOWNE TYPE G TIME LOCK

Yale's Type G time lock was introduced concurrently with Types D and E in 1888. (There is no record of a Yale Type F.) The Type G, which used two seventy-two-hour modular Waltham-made pocket watch movements, was Yale's smallest time lock to date. Its dial attributes a patent to Stockwell, but no Type G–specific patent has been identified.[18] Like the Types D and E, the two Type G movements each have an up/down power reserve indicator and are wound simultaneously by the single central key-driven arbor. Interestingly, American Waltham Watch Co. had specific names for each damascening pattern that they used on movements; the pattern used on the movements for Types B through G was designated "Hillside."

As with the Types D and E, production of the Type G ended in 1891 and was probably considered unsuccessful: only twenty-two Type Gs seem to have been made, with two consigned, but none known to be sold.[19] The example of Yale's Type G shown here controls a Yale No. 2 bolt motor, allowing the time lock to be used independently of any combination lock. It is the only known complete example of a Type G to survive. As shown by the heavy wooden case, this was a salesman's sample. It is believed that Yale mounted many if not all of their samples in similar cases, yet this is the only Yale time lock case to survive.

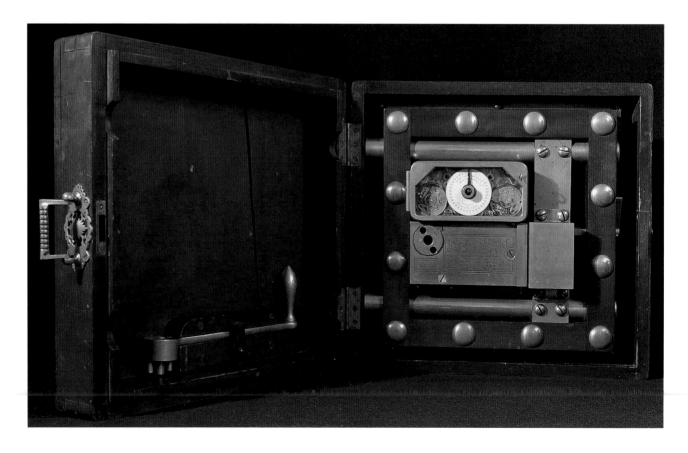

Yale's Type D offered the option of a key- and combination-free door, but it was large, requiring even more space than earlier models. For smaller safes, Yale introduced the Type G. With two Waltham movements wound from a single central arbor, the Type G offered the same benefits as the Type D, but in a more compact format. Seen here in a salesman's sample case, the Type G is mounted with Yale's more compact No. 2 bolt motor.

SARGENT & GREENLEAF'S TRIPLE A

Yale's 1888 introduction of the Type B and Type C time locks with their three semimodular movements was quickly recognized as a major advance, and in 1889 Sargent & Greenleaf began offering the Triple A, B, and C time locks, Sargent's first modular three-movement design. These Sargent Triples used three seventy-two-hour "L"-size movements in a jeweled bronze case but, unlike truly modular movements, they were position specific. On each movement, three small set pins on the interior case back align with holes in the movement's back plate, ensuring that only position-one movements fit into the first position, and so forth. The Triple A, B, and C were substantially similar in most respects, with the Triple A designed for use with an automatic, the Triple B with a bolt dog, and the Triple C with a bolt-engagement extension. A smaller-format Triple D entered production around 1896.

Each Sargent Triple employed a spring-loaded snubber bar that was nickel-plated and engraved. On the Triple A, the snubber bar runs along the case bottom to actuate an automatic—commonly the Burton-Harris automatic shown here—and was held in place by the snubber bar guide that bore the patent dates. This guide, located at the bottom left of the case, was possibly an antidynamite device, working with the adjustable tab and set pin under the center movement to keep the snubber bar from being pushed too far to the right or thrown out, off its mounting rail. The Triple A also featured winding eyelets through the rectangular door glass.

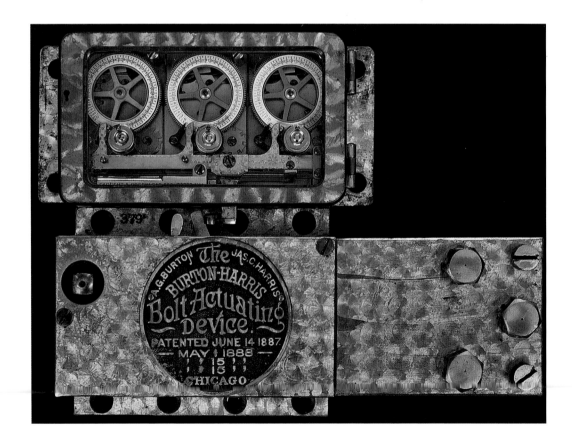

TRIPLE A

With the advent of ever larger vault doors with ever heavier boltworks at the end of the 1880s, the rise of the automatic bolt motor posed a problem for Sargent & Greenleaf, which had yet to offer a bottom-release time lock as a standard model. To work with the new bolt motors of large vaults and solid-door safes, Sargent introduced the Triple A, a time lock that offered not only a strong bottom release but also its first modular movements. The Triple A seen here is mounted to a Burton-Harris bolt actuating device, one of the first bolt motors to be used with Sargent time locks.

SARGENT & GREENLEAF'S TRIPLE B AND C TIME LOCKS

Sargent's Triple B included the low-profile version of its cello-bolt to dog the boltwork along the case bottom, moving the snubber bar to the case top. The tolerance between the dials and the snubber bar on these earliest Triple Bs was very small, making the lock more susceptible to dynamite, since an explosion could unseat the movements, which could in turn throw off the snubber bar and release the bolt. Unfortunately, the obvious solution of increasing this tolerance seems to have been foreclosed by Newbury's antidynamite patents, and Sargent's Triple B would remain unchanged until 1893. By that time the risk of patent litigation had all but disappeared and Sargent was free to increase the dial-to-snubber bar tolerance. The construction of the snubber bar on the Triple B also evolved, being cast in a single piece rather than having the extensions welded on, as they were in the earliest version.

Beginning around 1895, Sargent's third type of Triple featured truly modular movements for the first time. The

Sargent & Greenleaf adapted all the improvements of the Triple A for use in a side-release time lock, its Triple B. Sargent's new nomenclature (where the bottom release is now the "A" model and the bolt-blocking side release the "B") is of interest, possibly reflecting the growing popularity of automatics over manual bolts and marking the first time that Sargent's production suggests the primacy of the bottom-release format. This early Triple B shows the initial adaptation of the modular movements to the side-release bolt, with the snubber bar running above the movements rather than below as in the Triple A. Internally, Sargent's side-release bolt was still based on its 1877 cello-bolt, but no longer much resembled its namesake.

alignment pins were still included on the case back interior, but the patterns were now identical, allowing any post-1895 movement to serve in any position. This third version of the Triple B changed the most, with the snubber bar moving from the top to the bottom, like those found in the Triple A and C. With the repositioned snubber bar, the dial pin that actuated it had to be moved as well, now found in the unmarked area between the "0" and the "72" marks rather than at the thirty-four-hour mark.

After 1896, when "Sargent & Greenleaf Co." was added to the dial of the Model 2[13], changes to the Triples reflected those of the Model 2, except for a brief period around 1900 when the Triple's door glass took on a dual-scallop similar to the earliest Model 2s. Sargent's Triple A, B, and C were quite successful, and they were produced until 1922. While no detailed production records have been found for the Triple A or Triple B, case and movement serial numbering suggest that Sargent made at least one hundred each of the earliest Triple A and B, at least one hundred of the second version, and at least two hundred of the third. By the time production ended in 1922, total production seems to have reached over two thousand of Triple A and Triple B.

Detailed records for the Triple C do exist, at least in part because it seems to have been a special-order model, with only 150 Triple Cs sold from its 1889 inception until its 1922 demise along with the other Triples. These were delivered almost exclusively to the Damon Safe Co. of Boston through 1914, after which Damon disappears from the records entirely, and the balance of the Triple C went to the York Safe & Lock Co. of York, Pennsylvania.[20]

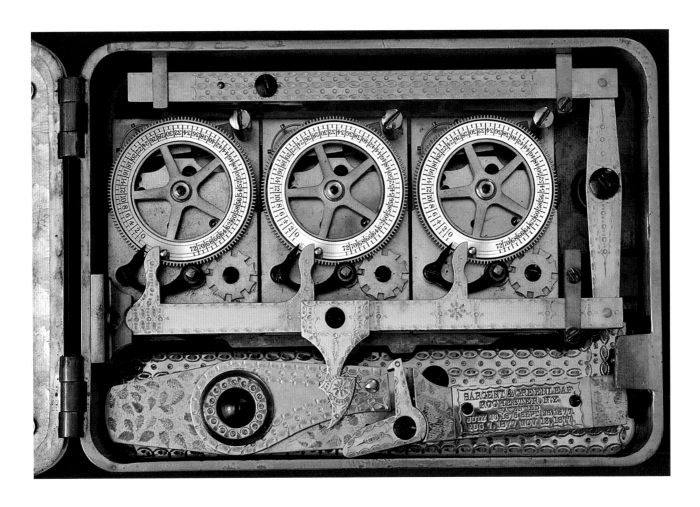

With Sargent's third version of the Triple B in 1895, the company had repositioned the snubber bar below the movements. Many subtler changes incorporated the antidynamite improvements originally patented by Newbury, an employee of Holmes Time Lock Co. By this time, however, there was a far lower risk of patent litigation, both across the industry in general and from Holmes in particular, who had been driven out of the time lock industry.

DIEBOLD TISCO SAFE PROTOTYPE TIME LOCK

The time lock shown here is listed in *Lure of the Lock* as simply a "Two Movement Time Lock" with the following description:

> There is no clue to the manufacturer of this lock, which was donated by the Mosler Safe Co., Hamilton, Ohio. There are two forty-six hour movements. 6½ × 2½ × 2 inches. Middle (shelf).[21]

Little has come to light about this time lock specifically in the intervening eighty-eight years, but it shares some interesting features with other time locks that hint at its source.

The time lock does not carry an attribution and is not a known production design, but during the 1890s the Diebold Company began manufacturing time locks for use in its Tisco line of safes and the Tisco-specific time locks were about the same size and shape as this one. Moreover, the Tisco-specific time locks featured a lever on the case bottom that could hold the release hook open, allowing the safe to be closed with the combination lock only during business hours. The same lever is visible here, to the left of the release hook. This time lock also has art nouveau acid etching on the movement cover, a style of case and safe plate decoration used by Diebold on its early time lock designs.

The movement format, however, is unique, with its two horizontally aligned escapements visible over the movement cover. A design so far afield from other time locks would normally be expected to be patented, but no patent for this time lock is known to have been granted. Consequently, this is thought to be a prototype or experimental design assembled by Diebold around the period in the 1890s when the company was developing its Tisco safe line but that never went into production.

DIEBOLD TISCO SAFE PROTOTYPE TIME LOCK

The John M. Mossman Lock Collection includes a number of locks whose history, without patent papers, production records, or litigation documents, can only be speculated on. This time lock's size, shape, decoration, and mechanism suggest that it was a design model or prototype made by the Diebold Company prior to beginning production of time locks for use in its Tisco safe line.

SARGENT & GREENLEAF BOTTOM-RELEASE MODEL 3 TIME LOCK

Beginning in the mid-1880s, a number of lock manufacturers began to devise automatic bolt motors, often referred to simply as "automatics." Wound with large wrenches, they housed extremely heavy springs that could draw back the increasingly complicated and weighty boltworks in the largest vault doors. One of the earliest bolt motor makers was Burton, Harris & Co. of Chicago. Originally, Burton designed its motors for use with time locks made by nearby Consolidated of Cincinnati. However, by 1890 Burton had developed a closer relationship with Sargent & Greenleaf, with Sargent eventually acquiring Burton at some point prior to 1901.[22]

When Sargent started working with Burton, Sargent had no time lock that could operate an automatic: even by 1890 most Sargent designs still had side-releasing bolts, whereas automatics generally require a bottom release. To develop a model for use with an automatic, Sargent's factory reworked an early Model 3. The case and door were cut and 1⅛ inches of the case were removed, disposing of the bolt and adding a slot in the case bottom through which the drop lever could directly actuate Burton's bolt motor. This operation also removed the door lock, which was replaced by a leaf-sprung clip screwed to the case exterior. The seam was delicately dovetailed and rejoined with exacting accuracy, and the surface was refinished. This model did go into production briefly as the Model 3A, using a one-piece cast-bronze case and the same seventy-two-hour movements as the Model 3.

BOTTOM-RELEASE MODEL 3

Special-order time locks were probably never common but may have been more numerous than would be expected. Although a small number of safe and lock makers did substantially dominate the bank security market, time lock makers seem to have been willing to accommodate older or less common safe formats. Further, true mass-production techniques were still a number of years off, making custom production only slightly more expensive. Close inspection of this Model 3 from Sargent & Greenleaf clearly shows that the entire bolt portion was removed, with the case bottom reattached with fine dovetailing, allowing the drop lever to act directly on the Burton-Harris automatic. Production records indicate that this style did see limited production as the Model 3A, but the success of Sargent's Triple A likely made it superfluous.

MOSLER SAFE & LOCK CO.'S CALENDAR TIME LOCK

Mosler Safe & Lock Co. had entered the time lock market by purchasing all of Beard & Bro.'s time lock interests around 1887,[23] and though Mosler was content for its first order with E. Howard to copy Beard's design, the company had also hired time lock engineer Phinneas King from Beard, expecting to develop its own improved design. By 1891 King had been awarded another patent[24] and Mosler was ready to introduce this model, the Mosler Calendar time lock.

Like the Beard design that preceded it, the Mosler Calendar used two seventy-two-hour E. Howard movements and kept the nickel-plated bronze case with the inverted "camelback" door shape. Because this example was made with a bronze door insert in lieu of glass, probably for use in a safe with coin traffic, it included two case-top windows to monitor the movements. The Calendar was also capable of unlocking and relocking during the course of the day, much like the Holmes Electric, but the major addition by Mosler was the return of King's calendar mechanism, first patented with Beard & Bro. in 1878 and appearing here for the first time with his latest improvements. The calendar mechanism is based around the seven-day dial, allowing the lock to be adjusted to pass over its daily open periods for as long as the movements have reserve power, allowing for Sundays, bank holidays, and other planned closures. According to King's 1891 patent, Mosler Safe & Lock envisioned the possibility of a calendar time lock with a thirty-one-day device, allowing the user to preset the holidays only once each month.[25] However, there is no indication that such a mechanism was ever made.

At $88.50 wholesale, the Mosler Calendar was an expensive time lock, possibly the most expensive two-movement time lock ever.[26] Mosler made an estimated two hundred of the model, but this is the only known surviving example. It has a bottom release for use with an automatic but, like earlier Mosler Safe & Lock time locks that were available with either a side or a bottom release, the Mosler Calendar was likely available with a side release as well.

When Mosler Safe & Lock Co. brought out the first update to its time lock in 1891, it was at once a major move forward and a return to original concepts. No longer sold as a low-cost alternative time lock, the Mosler Calendar included a number of features from Phinneas King's original 1878 patent that he had since improved upon, including a seven-day calendar mechanism, visible at the bottom right of the mechanism.

YALE & TOWNE TRIPLE L TIME LOCK

The rise in popularity among bankers of "solid" safe doors that relied only on a time lock and automatic bolt motor gave rise to a new generation of time locks designed for use with automatics. Yale & Towne introduced its Triple L time lock in 1892 to replace its earlier Type D and Type E models. Easily identifiable by its central bulge at the case bottom, the Triple L employed three seventy-two-hour modular "L"-size movements. This L-movement would go on to become the stalwart of the Yale line, used in a wide range of time locks.

Much like the Types D and E, the early Triple L time lock bore no attribution to Yale. The enameled dials have only numbers and the nickel-plated bronze case has only the open cross-hatch damascening common to Yale time locks but found on those of other makers as well. The example shown here with Yale's No. 1 Double Reserve Bolt Operating Device was sold to the German Illinois State Bank of Englewood, Illinois, on January 3, 1893.[27]

This first style of Triple L featured a large rectangular door-glass insert with winding eyelets through the glass. Beginning

YALE TRIPLE L

OPPOSITE: *As a model offering only a bottom release, the Yale Triple L was intended for use with an automatic bolt motor and worked well with the most technically advanced and reliable automatics of the day, such as Yale's Double No. 1, shown here.*

ABOVE: *The adoption of modular-movement design was a significant step forward for the time lock industry, allowing less-skilled technicians or even sales personnel to exchange malfunctioning movements and allowing repairs to be done by skilled professionals at a central location. Though the Sargent & Greenleaf Triple A of this period still had position-specific movements, Yale's Triple L featured truly interchangeable modular movements.*

with case serial number 1712, the Triple L's door glass was reduced to only the upper half of the door, placing the winding eyelets through the lower metal half to reduce glass cracking. Many other specialized variations of the Triple L were made to let this flexible design operate with almost any safe's boltwork. One such variation used a large bottom hook to control a combination lock, much like many of Consolidated's designs, and another moved the bulge opening to the back rather than the bottom. Some included an actuating arm extending from the left side of the case that was connected directly to the boltwork, making the bottom release unnecessary. In this configuration, the bottom bulge remained, but with no opening.

Yale also modified the Triple L with a Gesswein, or "G," Attachment, a secondary lock that prevented the time lock from securing the door during business hours. To lock the safe, the time lock would be wound and the plunger to the lower left of the bulge depressed. This freed the Triple L's snubber bar to slide to the left, moving the bottom release to the right, and securing the door. The Gesswein Attachment–equipped Triple Ls were most commonly used in cannonball-style safes, and often in Manganese Steel Safe Co. safes, on a mounting plate that would rotate with the door when locked. In this type of installation, the time lock would be at an angle whenever the door was open and was wound and maintained in this sideways position.

Many of the early L-movements for the Triple L were made by E. Howard. Contrary to some recent suggestions, Howard-made L-movements always carried an E. Howard attribution somewhere on the movement, whether on the dial, on the front plate obscured by the dial, or elsewhere. Other early L-movements were supplied by the Boston Clock Co. in 1893, as noted at the time by a technician in his service notebook (page 266). In 1894 Boston Clock Co. failed and Yale turned to the Seth Thomas Clark Co. for additional L-movements.

The easy replacement of modular movements makes dating any time lock by the movements' serial numbers inexact, but records show four clear production periods of the Triple L. The first, between 1892 and 1893, had movements serial numbered between 1 and 1000, supplied by E. Howard. These earliest L-movements had plain dials and their serial numbers engraved in cursive type at the bottom of the front plate.

A second period of Triple L production ran from 1892 until approximately 1894 and used L-movements made by the Bos-

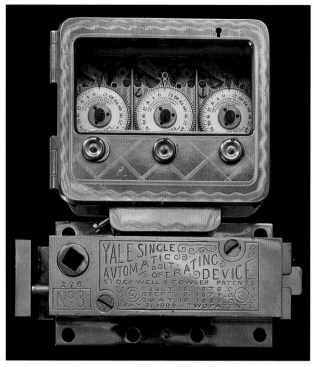

ABOVE AND OPPOSITE: *The Yale & Towne Mfg. Co. was among the leaders in automatic bolt motor design, offering a number of models, including the Single No. 3 seen here with the company's Triple L time lock. This Triple L features the Gesswein or "G" Attachment, visible just above the release with its actuating plunger extending out to the lower left through the bottom bulge. The G Attachment prevented unintended locking during the business day, a feature not seen since the Holmes New Electric time lock.*

ton Clock Co. serial numbered from 1001 to 2000. The subtle differences between escapements made by Howard and Boston Clock can be seen on page 267, but more obvious differences appear in the markings. The earlier movements in this range show the movements' serial numbers on both the dial and the plate bottom, while later ones have the serial number only on the plate and in minuscule type: "The Yale & Towne Mf'g Co." / "Stockwell's Patent" on the dial above the center and "July 19 1892" / "Stamford, Conn." below.

From 1894, Yale returned to E. Howard for L-movements, beginning the third period of Triple Ls that extended into 1895. Unlike the earliest plain Howard movements, these now show the movement serial number both in block numbers on the plate bottom and on the bottom of the dial, followed by "Stamford, Conn." The top half of the dial now

read "The Yale & Towne Mf'g Co." / "Stockwell's Patent." / "July 19th 1892." This third Triple L production run used movements serial numbered 3000 to 4500, leaving a large numbering gap between 2001 and 2999 that was never filled.

Around 1902 E. Howard stopped making time lock movements for the first time since 1875, a change associated with the takeover of Howard by the Keystone Watch Case Company.[28] The last period of seventy-two-hour L-movement production began in 1902 and continued until the 1920s, with Seth Thomas supplying all L-movements for Yale, numbered after 5000. Because these movements have no manufacturer's attribution, the possibility that they were made by E. Howard was not ruled out for many years. However, recent analysis of

Symbols and Serial No's of Yale Timelock Movements.

L. Howard	L. Boston	L. Howard
1 to 1,000	1,001 to 2,000	3,000 to 4,500

L. Thomas

From 5,001 to 7,435 have 3 move't scr's.
" 7,436 up " 4 " "
No's 7,403 and 7,409 " 4 " "
From 5,001 to 6,000 " large winding arbor pinions.

T. Thomas	M. Howard	M. Thomas
1 to	101 to 200	501 up 701 up have 4 scr's

List of Timelocks with new and old style glass

K. From 158 up have no eyelets in glass.
L. " 1711 " " " " " "
M. " 109 " " " " " "
N. " 202 " " " " " "
O. " 290 " " " " " "
P. " 151 " " " " " "

Never exchange Howard for Thomas mov'ts with auto's.

Y mov'ts # 366 up new style.

YALE TRIPLE L

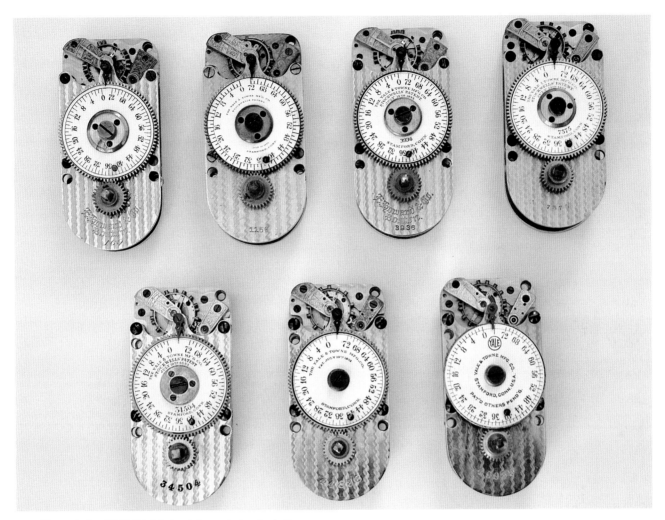

ABOVE: *The popularity of Yale's Triple L time lock meant a long production run, extending from 1892 until well into the 1920s during which more than sixteen thousand Triple L time locks were made. During this period the L-movement itself was made by a number of companies, with subtle differences in appearance. Top row from left: first E. Howard type, Boston Clock Co. type, second E. Howard type, first Seth Thomas type. Bottom row: second, third, and fourth Seth Thomas types.*

OPPOSITE: *Original documents such as this page from a Yale time lock technician's manual (complete with some additional notes from the owner) have shed new light on the numerous and shifting relationships between time lock makers and the clock and watch companies that made their time movements.*

surviving corporate records and subtle but notable differences between escapement designs clearly show these later, unmarked movements were made by Seth Thomas rather than by E. Howard.

The Triple L was one of Yale's most successful designs. Production eventually exceeded sixteen thousand time locks, making it the largest production run of any time lock of the era. Even today, the Triple L time lock is rather common and is found in many collections.

YALE & TOWNE TRIPLE K TIME LOCK

Concurrently with its introduction of the Triple L in 1892, Yale also introduced the Triple K. Based on a similar format to that of the Triple L, the Triple K also used three seventy-two-hour "L"-size movements but is distinguished by the extra space visible at the right side between the case and movements that houses the bolt dog.

Early on, the Triple K was a relatively minor variation and did not play an important role. However, the popularity of automatic bolt actuators began to fade between 1910 and the beginning of the First World War. With the return of hand-actuated safes, the Triple K became a major time lock model for Yale, and by the time production ended in the 1950s, more than three thousand Triple Ks had been sold.[29] Late in the model life, around 1915, some examples of the Triple K did away with the space in the right case interior, replacing it with an actuator arm. The example shown here was sold to the First National Bank of Juneau, Alaska, in December 1898.[30]

YALE TRIPLE K

The modular design of the Triple L offered sufficiently significant cost savings for Yale as well as convenience for the user that Yale introduced the Triple K, a side-release model for use without an automatic bolt motor. With its extra case space to the right of the movements for the bolt dogging mechanism in lieu of the bottom bulge of its more popular relative the Triple L, the Triple K was eventually quite successful in its own right.

CONSOLIDATED TIME LOCK CO.'S DALTON TRIPLE TIME LOCK

Milton Dalton was one of the most brilliant bank and time lock designers of all time and was awarded numerous patents that he assigned to Joseph Hall's companies. In 1893, Dalton patented what seems to have been his last contribution, the Concussion Triple. This magnum opus was a time lock whose patent alone spanned ninety drawings and thirty-seven pages of text making ninety-four patent claims.[31] The Concussion Triple was the last unique device in the development of time locks and marked both the culmination and the twilight of nonmodular time locks. From this point forward, the lower costs of maintaining and repairing modular movements would spell the end of built-in mechanisms.

What made the Concussion Triple unique was its anti-lockout device to release the bolt in the unlikely event the two movements failed. The mechanism is visible as a metal pendulum inside the left side of the case. To open the failed time lock, the safe would be rhythmically struck with a heavy ram, rocking the pendulum back and forth. Each swing of this pendulum would advance the time mechanism, eventually releasing the lock. This method of opening the Concussion Triple may seem long and noisy, but it was designed to be so. The mechanism guaranteed that the safe could be opened, but only after sufficient time and commotion, to ensure that attention would be drawn and to guarantee authorized use only.

ABOVE, LEFT AND RIGHT: *The final patent time lock known to have been designed by the brilliant Milton Dalton was his Concussion Triple, introduced in 1893. Although it shares a similar appearance with earlier Consolidated time locks, the Concussion Triple's internal workings were almost wholly new. The most notable change was the addition of the concussion mechanism, a metal mass visible to the left of the movements. Should the movements fail, the safe could be opened by rhythmically striking the outside with a ram, rocking the concussion weight and advancing the timers. (Courtesy of the Harry Miller Collection.)*

OPPOSITE: *This is but one of the twenty pages of drawings in Dalton's thirty-seven-page patent specification, undoubtedly the most complicated time lock ever designed. Unlike the final production design, the patent drawing shows a pair of concussion pendulums, one on either side of the case. Subtle and sometimes significant differences such as these between patent specifications and production were common.*

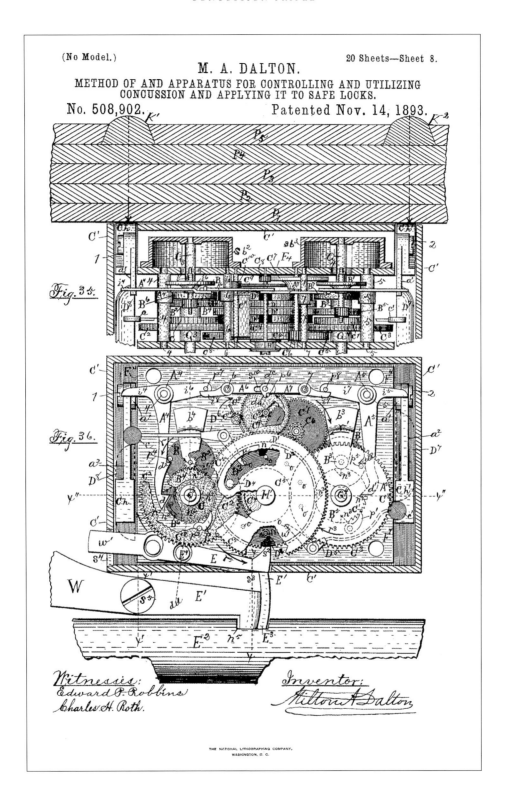

SARGENT & GREENLEAF'S MODEL O TIME LOCK

By the 1890s the time lock industry had reaped enormous benefits from the time lock's broad commercial adoption, the advent of modular movements, and a general expansion in the banking industry. These factors had led to time lock mechanisms that were exceptionally reliable and quite user-friendly—rare lockouts stemmed from user error, such as setting the time lock with only one operational movement. Concurrently, the architecture of the banking industry, of both buildings and the vaults they housed, had become symbols projecting stability, wealth, and power. These changes in banking led to a sharp growth in the number of extremely large vaults at the end of the nineteenth century. While not technically impenetrable, these vaults represented such great investment—and such staggering expense should they need to be forced open—that they required a time lock whose release was all but certain.

In response, Sargent & Greenleaf introduced its Model O time lock in 1893. Aside from its four seventy-two-hour L-movements[32] (identified by their curved, bottom-positioned indicator arrow and partial front plate that leaves the escapement clearly visible from the front), the Model O was fashioned much as a logical extension of other Sargent designs, with a jeweled bronze case, nickel-plated low-profile bolt, bottom-mounted snubber bar, square door glass, and the simple handcuff-type door lock now standard on Sargent's time locks.

The marginal benefits of a fourth movement represented a reasonable expense to only a small subset of the time lock

LEFT, TOP AND BOTTOM: *Two illustrations of bank vaults featuring four-movement time locks and automatic bolt motors by Diebold. As banks grew, their vaults grew larger and stronger, holding great amounts of assets for both the bank and their customers. Banks saw their vaults as a symbol of their strength and longevity and took great pride in them, often including vault images on their correspondence. This new stratum of vaults was so expensive (both to install and to force open) that they led to a new class of four-movement time locks from the major makers, seeking the last bit of certainty of proper operation.*

OPPOSITE, TOP AND BOTTOM: *Among the earliest four-movement time locks was Sargent & Greenleaf's Model O, introduced in 1893. The Model O did not offer much in the way of technical innovation, featuring four L-movements and a longer, heavier version of Sargent's two-piece bolt. The likelihood of proper operation was only marginally increased, since three working movements are all but certain to release the vault door. What the four-movement Model O did offer was the ability to schedule maintenance or repair of a movement while retaining the use of a three-movement time lock. In addition, the four-movement design was more proportionally sized to the large doors and could project the same strength and security as the vaults they were used in.*

MODEL O

market. One subjective benefit was the conspicuous investment in security—a large time lock proportional to a large vault door was synonymous with financial wherewithal. The extra movement did offer objective benefits that made these time locks appropriate for the largest vault doors, where the owners demanded absolute certainty of proper operation. However, rather than certainty of unlocking, the major benefit of the four-movement time lock was an assurance of regular operation, uninterrupted by maintenance or repair. Even with a malfunctioning movement, the Model O offered the industry-standard dependability of a three-movement time lock, allowing for confident use and a relaxed appointment for servicing or repair.

Due to the general rarity of the largest of vaults, the Model O was scarce even in its own day: fewer than three hundred are thought to have been made and fewer than ten are known to survive today. Sargent would go on to offer other four-movement models and other makers would design their own, yet the total market for this type of supreme security would always be small, making all four-movement time locks uncommon. Furthermore, the high profile that four-movement time locks operated under meant that almost every one was used extensively and maintained and repaired diligently, making any four-movement time lock with consecutively numbered movements exceptional.

Sargent & Greenleaf's 1893 introduction of its four-movement Model O time lock was, given its limited market, a commercial success. But the Model O was ultimately a three-movement time lock expanded to include an additional movement: the size of Sargent's L-movements and their associated release design gave them adequate strength for all common applications, but the new class of vaults emerging in the mid-1890s and their automatic bolt motors approached the outer limits of the L-movement's abilities. Yale, still only quasi-competitive with Sargent based on their 1877 interest pooling agreement, was offering both its Triple L with the Gesswein Attachment and its Quad N for unusually heavy release power. Sargent's Model O was a good option vis-à-vis Yale's Triple K, but the company would not offer a true counterpart to Yale's Quad N until its Model M time locks were introduced between 1901 and 1908.

The Sargent & Greenleaf M-movement[33] was larger than its L-size predecessor, with a small indicator arrow at the upper

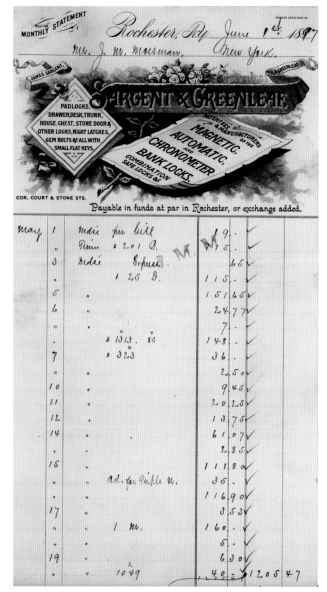

ABOVE AND OPPOSITE: *A monthly statement from Sargent & Greenleaf to John Mossman. Original documents such as these reveal the wholesale prices of various locks, parts, and services from their time, such as Sargent's four-movement time locks and repairs.*

left, a full front plate extending to the case top, and a small skeletonization behind the dial to monitor the balance wheel. More substantive differences between the L- and M-movements are less obvious, with the M-movement's construction more robust throughout, using heavier-duty escapements, stronger pivots, and an overall construction clearly proportional to its tougher intended use. Overall, the Model M was only about three quarters of an inch larger in height and width than the Model O, but this modest increase in size allowed for simpler, hence smaller, boltwork in the door.

YALE & TOWNE AUTOMATIC WITH HOLLAR'S TIME LOCK

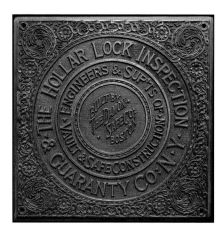

In the 1890s, with the advent of enormous main vault doors to project the prestige and security of banks, the boltworks and release mechanisms became proportionally larger, heavier, and more difficult to actuate. The Yale & Towne Mfg. Co. had for some years relied on its Triple K and L models, but in 1893 the firm introduced a more elegant solution, the Yale Quad N time lock.

The Quad N was a four-movement design, but unlike Sargent & Greenleaf's Model O (the other four-movement time lock then available) the Quad N used an entirely new movement format, the M-size modular movement. Originally made for Yale by E. Howard, the seventy-two-hour M-movement was a large-format design, with a broad face and wider, taller plates that allowed for a longer, more powerful throw—the combined pull of four M-movements exceeded seventy pounds.[34] Significantly overengineered with regard to its need for reliability, the M-movement was intended to be an impressive part of an impressive vault door.

Due to the thin market for four-movement time locks in general, E. Howard had made only two hundred of the M-movements before its exit from the time lock movement business. After this, all Yale movements including the M were supplied by Seth Thomas, with M-movement serial numbers beginning at 500.[35] Seth Thomas continued making the M-movement until about 1916 with the very last of these having a different dial design, including the Yale name in a circle at the top center. Almost all were seventy-two-hour movements, but a handful of the last M-movements could run for ninety-six hours. Only three of these ninety-six-hour Seth Thomas M-movements are known today. When the industry standard advanced to 120-hour movements after World War I, Yale offered to retrofit seventy-two-hour M-movements, a procedure that was still being done as late as 1970. Consequently, 120-hour M-movements can be found with any serial number and commonly display a ring of wear on the front plate around the winding arbor gear where the older, larger seventy-two-hour arbor gear was replaced with the smaller 120-hour gear.

The earliest Seth Thomas–made M-movements were used in a time lock made by a company owned by Hollar,[36] beginning in 1896, shown here. Based on the Yale Quad N design, the Hollar time lock included a patented device[37] that could be electrically triggered to rewind the time lock without opening the door. This modification involved cutting a round

ABOVE: *A bronze installation plaque for a vault built by Hollar and Damon, circa 1900.*

OPPOSITE AND PAGE 279: *The Hollar time lock took advantage of the redundancy of the fourth movement in Yale's Quad N design, replacing the third movement with an electric switch that allowed the user to rewind all three movements remotely without ever opening the safe. A review in* Scientific American *mentioned the usefulness in the event of a riot, an occurrence that bankers of the 1890s would have been aware of, having seen recent unrest from such issues as race and labor organization as well as earlier concerns such as the Civil War draft.*

hole in the case back for Hollar's large (over 4¼ inches in diameter) power reserve spring in a case and the replacement of the third movement with the electric switch that would be connected to the outside of the vault door through the connections on the case top. Other specially made parts on the Hollar time lock included the third through-glass eyelet for the oversized winding arbor, the angled snubber bar that could move clear of the electric winding motor, and the case itself, which was made by Yale and supplied only to Hollar with this polished bronze finish and machined scalloping.

A later, second type of the Hollar moved the rewinding switch to an area above the movements behind a plate, allowing the fourth movement to be reintroduced. This version was described in a *Scientific American* article, which explained the value of Hollar's device:

> Should conditions arise, however, which would, in the opinions of the proper custodians of the vault, justify them in keeping the vault locked for any additional number of hours, beyond the time for which it was originally set, this can be accomplished without opening the vault doors, and without anyone having access to the locks. The value of this feature may be illustrated when the contingency of fire or riot is considered, for in either case if would be undesirable to permit the action of the time-lock mechanism to make possible the unlocking of the vault. Under these conditions, all that would be necessary would be to simply close an electric switch, when the time-lock movements would be electrically rewound, thereby preventing the opening of the doors until the expiration of the added number of hours.[38]

Altogether, fewer than one hundred Hollar time locks are thought to have been made, and two of each type are known to survive today. This example was recovered following the demolition of an abandoned bank building in Indiana in 2002. The total production of Quad N time locks by Yale did not exceed four hundred, with fewer than five examples of the Yale Quad N remaining today.

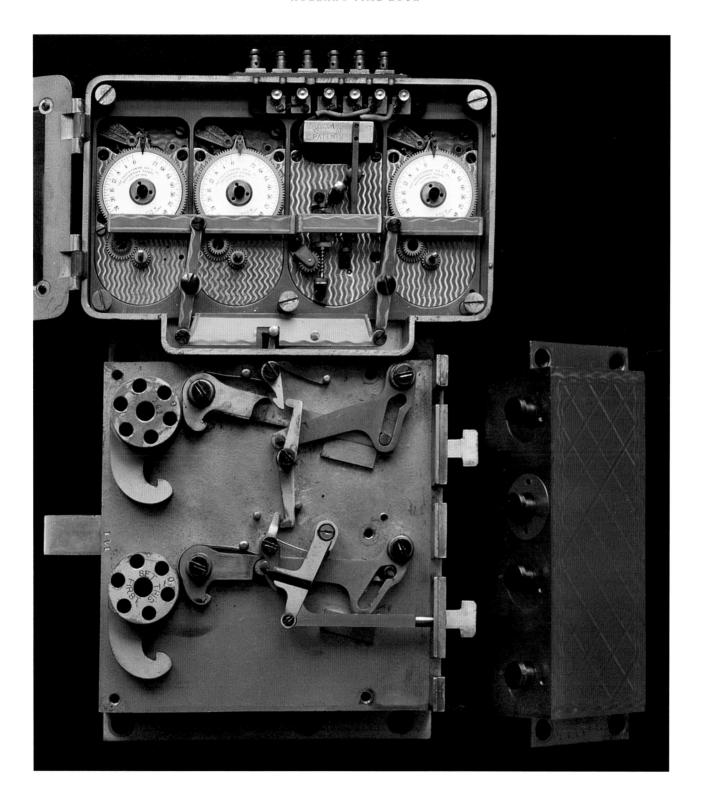

DIEBOLD SAFE & LOCK CO.'S TYPE 1 TIME LOCK

The Diebold Safe & Lock Co. first began operations around 1851 as a safe company, and it was not until 1894 that the company introduced its first time lock. Despite having waited so long to enter the time lock market, Diebold was a keen observer of other companies' time lock research and would go on to parlay its renown in safe making into confidence in its time locks. With three seventy-two-hour modular movements, the first Diebold model was the technical equal of any contemporary, offering unsurpassed reliability and ease of use for its time. The movements were very close in design to the Yale L-size movement and would be used by Diebold in a number of later models.

Like Yale's Triple L and K time locks made during this same period, early Diebold movements were supplied by E. Howard. However, E. Howard's watch-making portion was taken over by the Keystone Watch Case Company in 1902, and like Yale, Diebold switched to Seth Thomas for its movements. Diebold's pre-1902 E. Howard movements all bear an attribution to Howard, either under the winding arbor or behind the dial, whereas later Seth Thomas–made movements have no manufacturer's markings.

Diebold is thought to have made thousands of this first model and a few hundred still exist today. The example shown here is early, with E. Howard movements, intricate acid etching on the door, top, sides, and inner case plate, and gold plating on the bronze case. The attached Diebold automatic was patented and, like automatics made from Yale and Sargent & Greenleaf, came in various sizes.

ABOVE, RIGHT, AND OPPOSITE: *The Diebold Safe & Lock Co. produced a wide range of safes beginning in 1851, including the bank chests shown here. Both are "solid door" models that had no key or combination lock, rather relying entirely on Diebold's first time lock and automatic bolt motor. The round-door model is labeled on its foot as a Diebold Manganese Safe. The high percentage of manganese in this alloy produced an exceedingly tough steel that was not magnetic, an important quality for top security. (Courtesy of James Shoop.)*

DIEBOLD TYPE 1

DIEBOLD TYPE 1

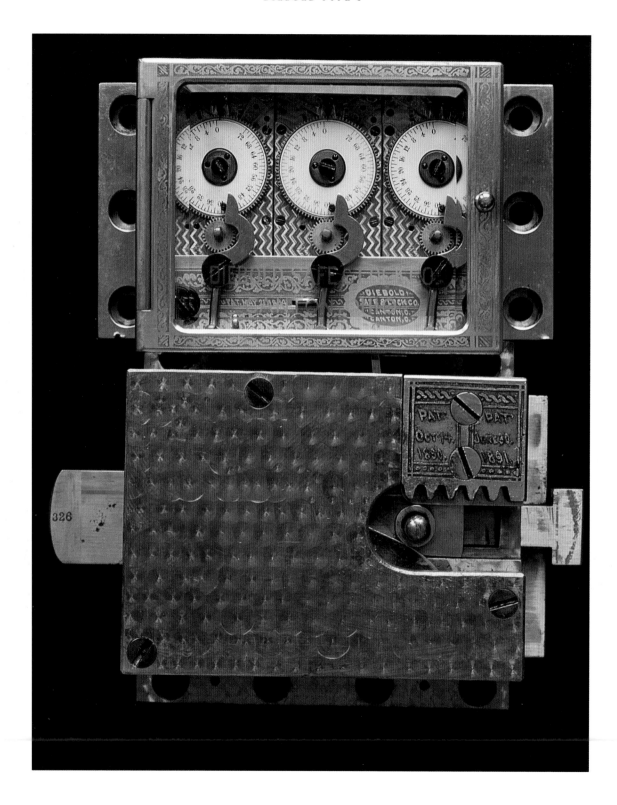

DIEBOLD TYPE 1

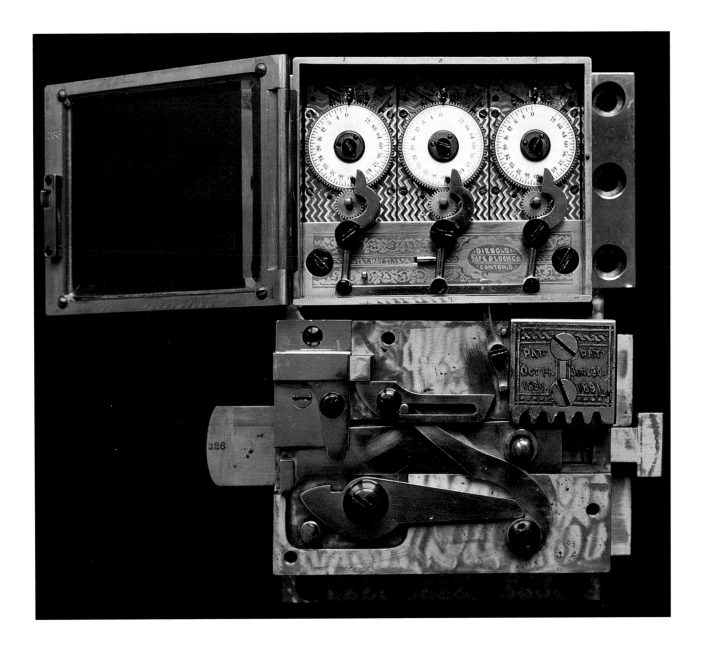

Diebold was a late entrant into the time lock industry but, as a well-respected safe and vault maker, the company could count on a market for its time locks while avoiding much of the litigation that plagued the industry early on. Diebold's first time lock appeared in 1894, a three-movement bottom-release model paired with its own automatic bolt motor. This model was used in a variety of safes and vaults, including the Diebold Manganese Safe on page 280 (left).

BLAKE'S BANK LOCK INSPECTION CO.'S COLUMBIAN TIME LOCK

The World's Columbian Exposition of 1893 was one of the great fin de siècle technical fairs that showcased the fruits of the industrial revolution. Along with the Exposition universelle de Paris of 1889, the Columbian Exposition offered a stage for innovators in all technical areas to introduce their products in a high-profile arena. One of these was Blake's Bank Lock Inspection Co. of Worcester, Massachusetts. Originally, F. H. Blake intended to compete with time lock manufacturers to provide time lock maintenance services, but he secured a series of patents in the early 1890s culminating in an 1894 patent[39] for what would be named Blake's Columbian time lock after its original venue.

Constructed under contract by E. Howard, the Columbian time lock used three seventy-two-hour modular movements notable for both their broad faces and the three-plate construction that added to the Columbian's strength and reliability, giving the movements enhanced stability and increased release power. However, these improvements were costly, pushing the price to $33.33 for each movement,[40] making them the most expensive modular movements made prior to World War I. Despite the expense, Blake's Columbian was successful, offering the prestigious large-format movements found on even more expensive four-movement time locks, such as Yale's 1893 Quad N, in a three-movement design.

The Columbian's nickel-plated bronze case set it apart from its contemporaries in both appearance and function. The case is plain except for the door, which featured an unusually elaborate art nouveau design applied via acid etching. Blake also included a unique reversible bolt mechanism. As claimed in his 1894 patent, the boltwork connector, shown here installed to the right, could be exchanged with the threaded case plug from the opposite side, allowing the one design to engage the safe boltwork from either the left or the right. Blake's reversible bolt would be adopted, altered, and independently patented by many other makers, eventually to become a design standard. Unfortunately for Blake, he would receive neither royalties nor recognition during his time.

Blake ordered at least seventy-seven movements from E. Howard between 1893 and 1897, enough for twenty-five time locks.[41] It is not known just how many of these were assembled, but the example shown here is the only one known to remain today.

Blake's Columbian time lock was named for the World's Columbian Exposition of 1893, where it made its debut. Expensive to produce due to its unusually large movements and intricate acid-etched decoration, Blake's Columbian sought to justify its higher price with its ability to project a powerful and prestigious appearance, yet it was not without innovation. The bolt connection visible at the top right of the case is reversible, a feature that would quickly turn up on other companies' time locks. Unfortunately, no benefit from this clever improvement seems to have made its way back to Blake.

SARGENT & GREENLEAF'S CARY SAFE TIME LOCK

The Cary Safe Company of Rochester, New York, was among a fair number of safe makers offering time locked freestanding money chests for bank use around the end of the nineteenth century. Many of the more successful safe makers developed a close relationship with a specific bank lock maker, and Cary Safe had such a rapport with nearby Sargent & Greenleaf that allowed Cary Safe to offer its own custom-built time lock in its patented screw door model safes,[42] beginning around 1893.

The only known surviving example of the extraordinary Cary Safe time lock includes the case and automatic bolt motor but lacks the movement. The case door indicates that the time lock was patented and, indeed, two time lock patents were awarded to Cary employee George Goehler that included the automatic. However, neither shows how the automatic worked with the time movements. These movements were most likely a pair of seventy-two-hour movements made by Sargent, but no example has been found.

The case is bronze with a scalloped bottom that is fitted to Cary's screw door. Rather than a door-glass insert, this time lock features a singularly intricate art nouveau intaglio design. The inner side of the door has a jeweled indent that corresponds to the outer design, which suggests that this example may have been intended as a coin safe. This is the only known example, but it is believed that a number of them were made.

The Cary Safe time lock case seen here is the only known surviving part of this example of Sargent & Greenleaf's close operations with Rochester cohabitant Cary Safe Company. The rounded bottom fit the curve of the screw-type door.

DIEBOLD SAFE & LOCK CO.'S TISCO TIME LOCK

Beginning in 1895 and continuing through 1900, the Diebold Safe & Lock Co. produced a number of time locks designed specifically for use with its Tisco line of manganese steel bank safes. This time lock used two seventy-two-hour E. Howard movements to control a Diebold-patented safe lock, first introduced in 1885. The safe lock featured four tumblers, an offset spindle, and was marketed as "anti-dynamite." This claim was not hollow, as Diebold spent significant effort during the production period to test both the lock and the Tisco safe with increasingly strong explosive charges. This series of tests culminated in 1910 with a test explosion of ten ounces of nitroglycerin, which the safe and lock withstood.

Unique to certain Tisco safes, the time lock is mounted upside down by design, with the case door opening from the top down and the movements releasing when zero hours reaches the bottom of the dial. This may have minimized mounting space in the tight area of the Tisco door. More than twenty-five examples of this time lock and a few hundred examples of the safe lock survive today. The example shown is the only one known with the time lock, safe lock, boltwork, and patented offset spindle in the original mounting frame.

TISCO TIME LOCK

OPPOSITE AND THIS PAGE: *In a wonderful series of photos, we see the extent to which Diebold went to better understand the explosive resistance of its Tisco safe. Conducted in 1910, the test was well documented, with testing teams from Diebold and the Marietta Torpedo Company in attendance. The testers removed the hinges and then attempted to blow off the locked door with amounts of nitroglycerin ranging from an eighth of an ounce to ten ounces. After more than five hours and sixteen detonations, the safe was still secure and the Marietta Torpedo Company certified further attempts useless. (Courtesy of James Shoop.)*

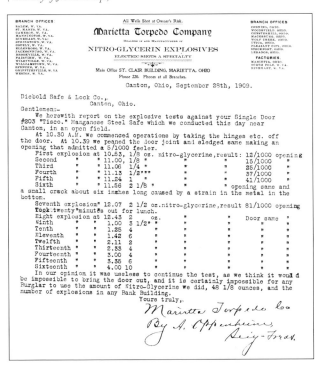

Diebold's line of Tisco safes featured nonmagnetic manganese steel construction, a top-quality Diebold lock, and a compact time lock specially designed to fit inside the Tisco's door frame. Though this time lock is not known to have been used in other safes, it was properly mounted upside down in the Tisco, with its door opening down and unlocking when the "0" reached the bottom of the dial.

DIEBOLD SAFE & LOCK CO.'S FOUR-MOVEMENT TIME LOCK

Sargent & Greenleaf's 1893 introduction of the four-movement time lock was only a limited threat to Yale, given its popular Gesswein device–equipped Triple L time lock as well as the interest pooling agreement with Sargent. However, it was a major threat to Diebold, a company that had only recently managed to gain a foothold in the high-end time lock market. Faced with this major move by Sargent, many other smaller companies simply did nothing, having no aspirations to

Diebold was a major installer of some of the largest vaults in the banking industry, such as the safe deposit vault shown here. "Diebold Safe & Lock Co." can be made out engraved on the glass that covers the central area of the door where the time lock and two combination locks are mounted. The enormous hinges support the great weight of the door but are not part of the vault's security, thanks to a boltwork that secures the entire circumference of the round door. (Courtesy of Jim Shoop.)

DIEBOLD FOUR-MOVEMENT

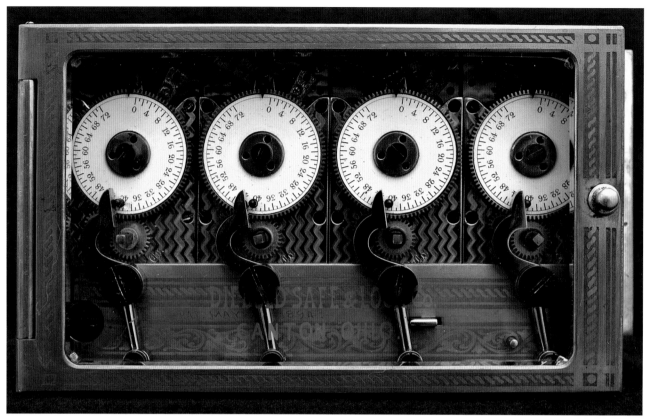

ABOVE AND OPPOSITE: *The introduction of four-movement time locks in the 1890s was simply a nonissue for many smaller companies that could not compete at the highest level of vault installation, but for Diebold this was a significant part of its business. Being the designer, producer, and installer of the vault, lock, and time lock meant that Diebold likely had little need for increased release power or reliability to operate its vault doors, since all parts were designed to work together. However, four movements was quickly becoming the industry standard for flagship time locks and Diebold obliged, extending its successful three-movement model to include a fourth movement.*

compete at the most expensive levels. But Diebold was in the business of constructing and installing top-line vaults and may have been unwilling to let its inside position be swept up by Sargent.

Consequently, soon after the earliest four-movement time locks appeared, Diebold adapted a four-movement model of its already successful three-movement format. With its four seventy-two-hour E. Howard–made movements and ornate acid-etched bronze door, top, and sides, this model offered unsurpassed reliability and projected stability as well as any Yale or Sargent & Greenleaf four-movement model. And though Diebold's four-movement did not offer the increased release power of Sargent's Model O or Yale's Model M, this was of little concern to Diebold, which installed most of these time locks in vaults of its own construction, allowing the company to monitor release requirements in a way that Yale and Sargent could not.

Based on an 1894 patent by Kirks,[43] the earliest Diebold four-movement time lock was available only with a bottom release for an automatic. (When the use of the automatics fell out of favor at some point in the 1920s, Diebold adapted this design again to operate with a bolt-dogging side release.) The bottom release operates behind the acid-etched plate below the movements, with the snubber bar serving only to coordinate all the actuator arms to the third, which pushes the small nickel-plated release stub. Diebold is thought to have pro-

292

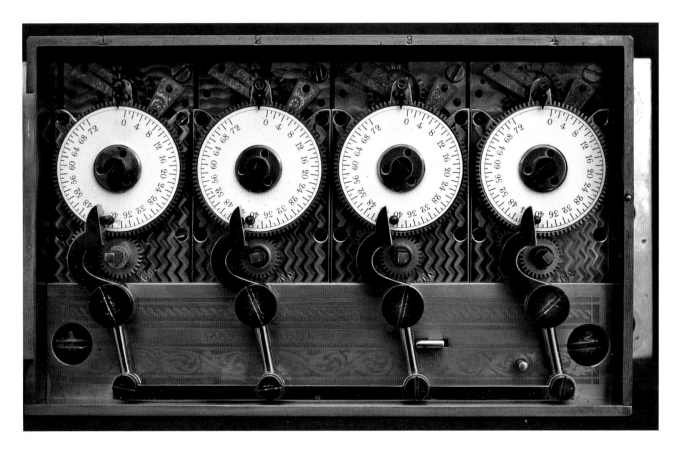

duced a few hundred of its four-movement time locks, but likely fewer than Yale or Sargent, making it quite uncommon, even among four-movement time locks. Five examples of Diebold's four-movement with acid etching are known to exist today as well as two examples with the satin bronze finish introduced after 1900.

YALE & TOWNE QUAD M TIME LOCK

Yale & Towne Mfg. Co.'s Quad N time lock was, for six years, the largest, most powerful time lock available anywhere. In December of 1899, however, Yale introduced the Quad M, featuring the same four seventy-two-hour M-movements as the Quad N, but designed with a side release rather than the Quad N's bottom release for an automatic. With its extended case to receive a bolt extension, the Quad M was the largest and heaviest time lock ever made, and like the Quad N the combined pull of the four movements exceeded seventy pounds.[44] This was an odd bit of advertising because, unlike the Quad N, the Quad M mechanism did not depend on the movements to "pull" anything. The Quad M was sold

for use with hand-operated boltworks only and simply dropped the side-release block in the case.

The Quad M included an "Open Attachment" visible through the case door to the side of the movements. First found on the Holmes Electric time lock, introduced in the late 1870s, this mechanism allowed the time lock to be wound and set but not actuated when the vault door is locked with the combination lock during the business day. Once the Open lever is thrown up, the time lock will secure the door when it is closed and locked. The Quad M could be made with the bolt opening at the top left or right, but an "M Special" was also available with the opening at the bottom left or right.[45]

The nickel-plated example seen here was sold to the People's Savings Bank in Bridgeport, Connecticut, in 1909,[46] but by this time the nickel-plated finish was available by special order only.[47] From 1908, Yale began a transition to a bronze case with a textured surface, using a pattern of overlapping circles. Of the earlier type of Quad M, identified by winding eyelets through the door glass, Yale made 109 and two are known to survive. Later production featured eyelets through the metal lower half of the door. Total production numbers are not known with confidence, but fewer than four hundred are thought to have been made and fewer than twenty-five remain today.

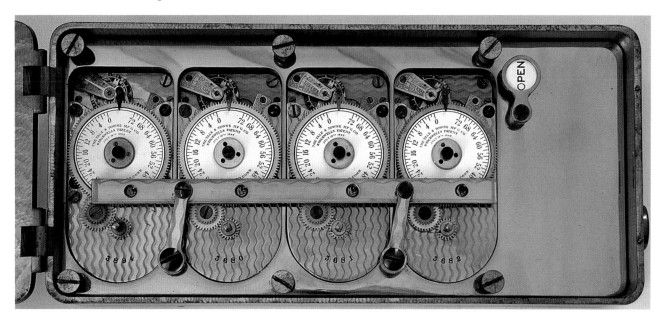

OPPOSITE, TOP AND BOTTOM: *The Quad M time lock from Yale was one of the most successful four-movement designs, with more than three hundred thought to have been produced. While such production figures pale in comparison to models such as the company's Triple L, the potential market for the Quad M was much smaller, being limited to the largest vaults with hand-operated bolts.*

ABOVE, TOP AND BOTTOM: *The operating mechanisms of major time lock models changed little in the early twentieth century, although time lock finishes and appearances did change based on design trends. The Quad M Special (bolt release on the bottom right) shown here was made after 1908 when Yale introduced a new and possibly less expensive surface, replacing the damascened nickel plating with a textured bronze surface.*

CHAPTER 6

Locks of the Early Twentieth Century: 1900–1915

At the start of the twentieth century, banking and finance in the United States had come into its own as a major industry, as had bank security. Thanks to the combination and time locks of earlier periods, stealthy burglary of bank safes and vaults had given way to far less frequent but much more hazardous daylight robberies on the banking floor, in the style associated with the "wild west" and Bonnie and Clyde. Consequently, the development of new bank locks started to slow as makers began to focus on the ease of manufacture, the ease of maintenance, and on new formats. The new models introduced during this period were, however, no less intricate or beautiful than those of the previous century.

CHICAGO TIME LOCK CO.'S MARSH TIME LOCK

Beginning soon after 1900, the Chicago Time Lock Co. debuted the first production time lock that offered a ninety-six-hour power reserve. Based on a design for which Earnest Marsh would eventually be awarded a patent,[1] the earliest style of this time lock had a nickel-plated case housing three movements with both a larger twenty-four-hour primary dial and a smaller secondary dial above, numbered with only 0, 24, 48, 72, and 96.

This first version was soon replaced by a second style (seen here) that included a number of revisions to substantially finalize the format. The secondary dials were done away with and the main dial, now marked through ninety-six hours, was repositioned at the top of the movement. The case edges became rounded and the finish was polished bronze, although nickel plating may have been available. This second model also introduced an optional mechanism to hold the wound time lock open during business hours, similar in function to the disabling latch on the Holmes Electric, but controlled by a fourth arbor visible at the case bottom between the second and third movements' snubber bar pins.

A third version of the Marsh time lock updated the case for use in a cannonball safe, adding a half-metal door with eyelets for the winding arbors and the fourth hole arbor if present. Available in both textured bronze and acid-etched nickel plate, this third style of the Marsh time lock would be Chicago Time Lock's last before apparently being taken over by Diebold around 1908. Around this time, Chicago Time Lock production ends and Diebold begins offering models with only a change in dial markings from "Chicago Time Lock Company./Chicago, Ill./U.S.A." to "Diebold Safe & Lock Company/Canton, Ohio/U.S.A." All parts on these time locks are interchangeable.

The ninety-six-hour movements in the Marsh time lock do not bear a maker's attribution but are thought to have been made by the Hampden Watch Co., or possibly by its parent, Deuber Watch Co., both of Canton, Ohio. The design of and engraving on the escapements is consistent with movements made by these companies at this time.[2] Production records for Chicago Time Lock are not known to survive, and only five examples of its ninety-six-hour model are known to survive today: one of the first and third styles and two of this second style.

MARSH TIME LOCK

The Marsh time lock, the first to offer a ninety-six-hour power reserve, was made by the Chicago Time Lock Co. between 1900 and 1908, before Chicago Time Lock was likely taken over by Diebold. On this example of the second version, the arbor for the optional disabling mechanism can be seen on the bottom plate between the second and third movements' snubber bar pins.

YALE & TOWNE K31½ TIME LOCK

Among the safe companies with sufficient market presence to offer a specialized time lock was the New York City–based Hibbard-Rodman-Ely Safe Co., which, beginning in 1902, made a freestanding bank chest featuring a two-case time lock system from Yale. Denominated the Model K31½, each case included two Seth Thomas–made seventy-two-hour movements installed in sequentially numbered pairs. However, each K31½ housed both the two-movement time lock and a four-tumbler combination lock. The example shown here was made early in the production run, in November 1903, and installed in a safe for the Farmer' and Merchants' Bank of Spokane, Washington, on March 2, 1904.[3]

Yale produced the 2 x 2–movement K31½ for Hibbard-Rodman-Ely during a span of only about two years into 1904. By that time, Hibbard-Rodman-Ely seems to have sold all its operations to the Manganese Steel Safe Co.,[4] and Yale continued to make the K31½ for Manganese Steel Safe until at least 1911[5] and possibly as late as 1916. Total production of the K31½ certainly reached fifteen hundred time locks, and possibly two thousand,[6] enough for seven hundred and fifty to one thousand safes. Today, fewer than one hundred individual K31½ time locks survive with fewer than twenty in sequentially numbered pairs.

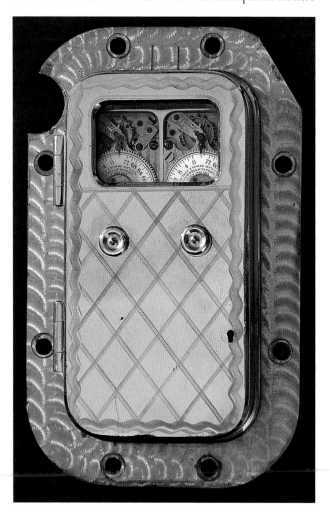

Despite the emergence of certain time locks as market leaders (such as Yale's Triple L), makers did introduce new designs, commonly in conjunction with a major safe maker. Such a model was Yale's K31½, designed specifically for use with Hibbard-Rodman-Ely bank chests. Behind the release mechanism below the movements, the K31½ housed the wheel pack of a four-tumbler combination lock, making each K31½ a small single-case bank lock in the genre of the long-defunct Sargent Model 1 and Dalton Triple Guard. However, rather than being a large flagship model, the K31½ was compact and functional, intended to be installed in pairs.

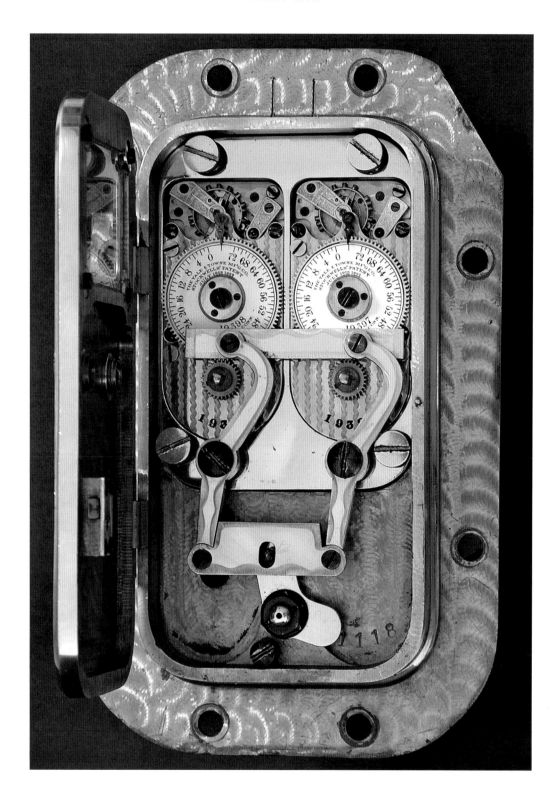

SARGENT & GREENLEAF'S MODEL 6 TIME LOCK

From its 1874 introduction of its Model 2, Sargent & Greenleaf sought to expand its market by adapting the basic Model 2 movement to ever smaller formats. By 1900, however, Sargent's trademark cello-bolt had been whittled down to minimum dimensions and it seemed that this reducing trend would come to a halt with the Model 4. In order to reduce size requirements further, Sargent would have to make a dramatic departure from what had been possibly the most successful time lock bolt design ever. The result was the Model 6, a small two-movement seventy-two-hour time lock and the first bolt-dogging Sargent time lock to abandon the cello-bolt since its inception in 1877.

The new bolt was moved to the top of the case behind the movements, with a coil spring that forced it up to the locked position after the door's boltwork is thrown shut. This required that the movements be wound and the bolt be cocked prior to locking. The cocking mechanism evolved over the production life of the Model 6, beginning as a key-turned arbor located between the movements at the center top of the case. This arbor was quickly replaced by a simpler button inside the case and finally grew to the two-lobe button that could be finger-set with the door open or closed, as seen here.

Detail surrounding Sargent's Model 6 production is not as well known or documented as other models, but case and movement serial numbering suggest that it was first seen between 1900 and 1903,[7] and was available until at least 1929 in three different sizes.[8] Although total production is estimated to have eventually exceeded two thousand, fewer than seventy-five examples of all styles of Sargent & Greenleaf's Model 6 are estimated to survive today.

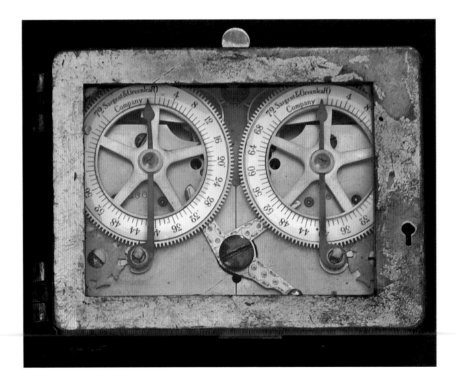

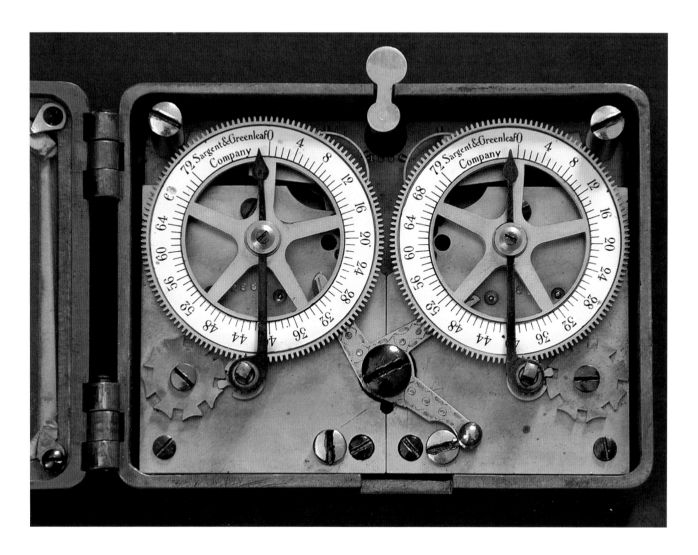

Sargent & Greenleaf's Model 6 was the company's first production design to offer an alternative to both its bottom-release and cello-bolt models. This new bolt was located behind the movements at the top of the case and was set by the user with one of three cocking methods, the last and best being the two-lobed button seen here. After winding the movements, the button was pushed in, loading a coil spring that would block the bolt once the door was shut. This new device included the benefits of Yale's Triple L Gesswein Attachment and could be used by someone without access to the time lock, thanks to the button extension outside the case. Of course, this last bit of security was little more than convenience, due to the glass door and weak handcuff-style lock.

CONSOLIDATED TIME LOCK CO.'S
HARRY DALTON AUTOMATIC TIME LOCK

In the early 1900s, the Ely Norris Safe Co., operating in Perth Amboy, New Jersey, introduced a new solid-door model of cannonball safe. The company then looked to the Consolidated Time Lock Co. for a time lock to fit the round safe door. As the Ely Norris design was intended as a commercial money chest, it would have no combination or key lock but rather would be open during all business hours, a common feature of the time.

Consolidated produced a time lock design that featured a round case with an internal automatic bolt motor controlled by three movements that were based on a new patent awarded to Harry Dalton[9] in 1904.[10] These movements had a seventy-eight-hour power reserve[11] that at first were made by the Elgin Watch Co. with white enamel. At some unidentified point, Consolidated's movement construction shifted from Elgin to the South Bend Watch Co., identifiable by black enamel dials. The movements were wound through eyelets in the glass, and the automatic through the eyelet at the case top. These winding eyelets were more important to this time lock than most earlier designs, since the nickel-plated bronze case has no door. Rather, its lid is held on by two set screws at the ten o'clock and four o'clock positions through the case side. The example shown here retains the complete working boltwork, including the triggering lever, running along the left door bolt. With the time locks and automatic wound, the safe door would be closed and rotated into place, depressing the lever and shooting the bolts out. Today, three examples of the Dalton Consolidated with the internal bolt motor time lock are known to remain.

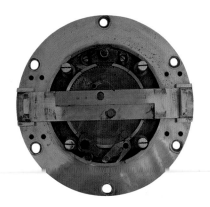

ABOVE AND RIGHT: *Joseph Hall had created Consolidated to protect his Hall's Safe & Lock Co., yet it was Consolidated that survived under Hall family ownership long after Hall's Safe & Lock had been sold. Consolidated Time Lock Co. produced this Harry Dalton patent design for the Ely Norris Safe Co. beginning in 1904. Designed to fit closely into the mounting space in the door, this time lock left no room for any other mechanism. Consequently, it operated without a key or combination lock and incorporated its own light automatic inside the case behind the movements. The example seen here is complete with its boltwork and actuating armature.*

OPPOSITE: *The low-profile automatic seen here was an important part of Harry Dalton's design. Shown here both unlocked (far left) and locked (center), this bolt motor allowed Ely Norris to include both the time lock and automatic in the space normally allocated to a time lock only. Later models used a lever at the mechanism's top (far right) in place of a winding arbor, allowing the user simply to slide the lever to the right to cock the automatic.*

CONSOLIDATED TIME LOCK CO.'S
HARRY DALTON TRIPLE TIME LOCK

With Consolidated Time Lock Co. suffering sharp competition from midwestern competitor Diebold, innovation may have taken a backseat to corporate survival. Harry Dalton's movement design that Consolidated used for its Ely Norris Safe Co. time lock proved to be so well received that it warranted adaptation for a broader market. Using the same seventy-eight-hour Elgin Watch Co.–made movements, Consolidated assembled a general-purpose time lock in a standard nickel-plated bronze case with a windowpane damascening pattern on the door and diamond crosshatch on the sides and top.

The example of Consolidated's Harry Dalton time lock shown here was the first and most common style, with three movements and a bottom release for use with an automatic. However, Consolidated also made two- and four-movement versions and offered a bolt-dog adapter for use without an automatic. This adapter was a separate thin steel mechanism attached to the case bottom, containing a steel disc with a hole that the snubber bar would rotate into alignment with the bolt. This example includes the original Elgin movements, whose serial numbers place it among the earliest made.[12]

HARRY DALTON TRIPLE

The success of the Harry Dalton patent time lock with Ely Norris led Consolidated to adapt the design for broader use. The round door insert and the internal automatic bolt motor were done away with and replaced with a more common nickel-plated bronze case and bottom release. Although production records for Consolidated's Harry Dalton Triple are not known to survive, this must have been reasonably successful, since it is known also to have been available in two- and four-movement models.

YALE & TOWNE DUPLEX TIME AND COMBINATION LOCK

Along with its assumption of Hibbard-Rodman-Ely Safe Co.'s contract for Yale's Model K31½ time lock in 1904, the Manganese Steel Safe Co. began offering a similar, smaller two-case time lock designated the LS31 but advertised as the Yale Duplex Time and Combination Lock.

As with the K31½, each LS31 controlled its own internal four-tumbler combination lock housed in a bronze tub. However, with only a single seventy-two-hour Seth Thomas L- movement to release each single-custody lock, Yale's instructions for the LS31 made clear that if a movement stops during winding, the user was to remove that time movement with some small mechanical talent.[13] (Closing the safe with the malfunctioning movement left in would leave that lock permanently dogged, and with only a single movement to release the other combination lock the risk of a lockout was always too great.) The resulting interim mechanism had one time-locked combination and another that was not time locked—not ideally secure, but a useful compromise until the modular movement could be replaced, likely the next business day.

Between the beginning of production in 1904 and 1911, Yale made more than fifteen hundred of the LS31, enough for seven hundred and fifty safes.[14] However, total production is certainly greater, with Manganese Steel Safe continuing to make its bank chests until about 1916 and continuing to use Yale's Duplex time and combination lock for at lease part of that period. Fewer than three hundred individual LS31 mechanisms are thought to survive today, with fewer than one hundred of those in consecutively numbered pairs.

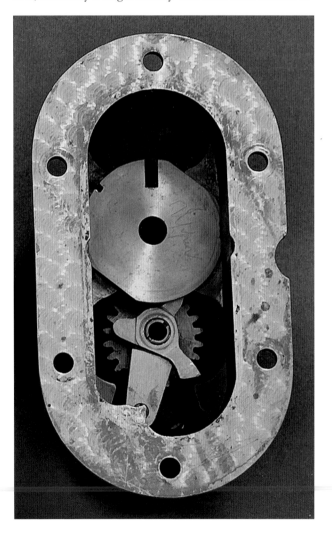

The Yale Duplex time and combination lock, referred to in records as the LS31. Similar in concept to the K31½, the LS31 was sold and installed in pairs, each unit featuring both a time lock and the wheel pack of a four-tumbler combination lock. Unlike the K31½, the LS31 time locks had only a single movement, requiring some method for disabling the time lock should one movement fail. Yale's solution was to make the time lock's case removable from the bronze shell that held the wheel pack. Should a time movement stop, the user would simply remove that movement until it could be repaired.

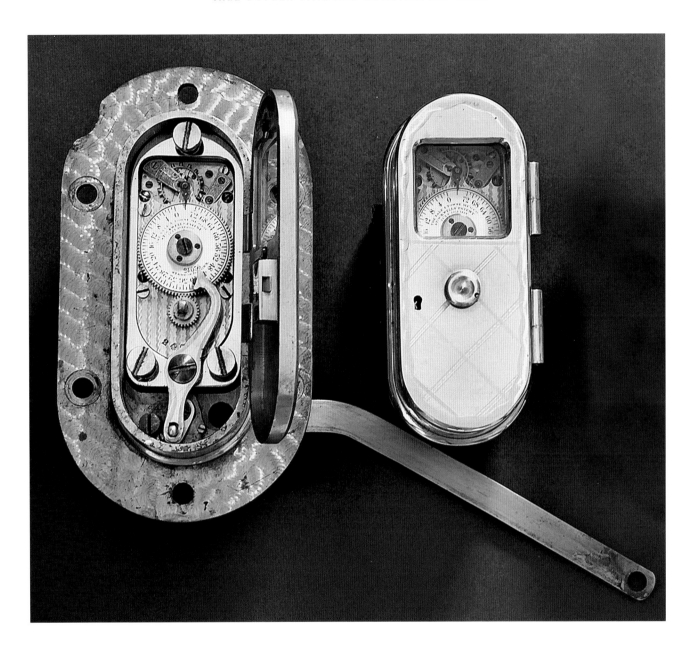

YALE & TOWNE T-MOVEMENT AUTOMATIC TIME LOCK

At some point during 1906, the Consolidated Time Lock Co. stopped production, bringing to a close the last of Joseph Hall's legacy of independent lock production. While historically interesting, this also had commercial consequences. One of these was to force the Ely Norris Safe Co. to go elsewhere for time locks for its cannonball-style safe now that Consolidated's Dalton-designed round time lock was no longer available.

Ely Norris went to Yale, which began production of its T-Movement time lock in 1907 or 1908. The earliest Yale time lock to replace the Consolidated model may have incorporated Yale movements in a steel block made as a retrofit for existing Consolidated cases, but only one poor-quality copy of a photograph of this has ever been found.[15] However, Yale had no long-term need for Consolidated's case design and by 1908 was making its T-Movement time lock completely in-house. As with its precursor, Yale's T-Movement used three seventy-two-hour modular movements in a round case that also housed an automatic bolt motor, independently wound from the arbor at the case top. The case featured the same concentric-circle machine engraving as the Consolidated model it replaced, but Yale's boltwork connected vertically within the back of the case rather than from the sides. Later production of the T-movement mounted the three movements in a staggered format, with the center movement raised into the empty space of the round case and the automatic mechanism moved to the increased area below.

The case serial number for a Yale T-Movement time lock is found on the case interior, behind the movement mounting block. (This example is serial numbered 17.) The visible number "518" on the mounting flange was likely added by Ely Norris, corresponding to the safe serial number that the lock was installed in. Production of both the early and the later types of the T-Movement time lock are thought to have been quite limited, ending with Yale's introduction of the Model 361 in 1908. Today two examples of Yale's earlier T-Movement time lock and three of the later, staggered T-Movement time lock are known to survive.

The Yale T-movement time lock seems to have arisen in response to a vacancy in the market after Consolidated Time Lock disappeared in 1906. Like Consolidated's Harry Dalton patent, the T-Movement time locks featured a round case and an internal automatic behind the movements. Unlike Dalton's patent, however, it used a vertical bolt that connected to the door with extensions through the top and bottom of the case. The lid had no hinge and was generally expected to be left in place, with the winding arbors for all three movements and the automatic accessible through eyelets.

YALE T-MOVEMENT

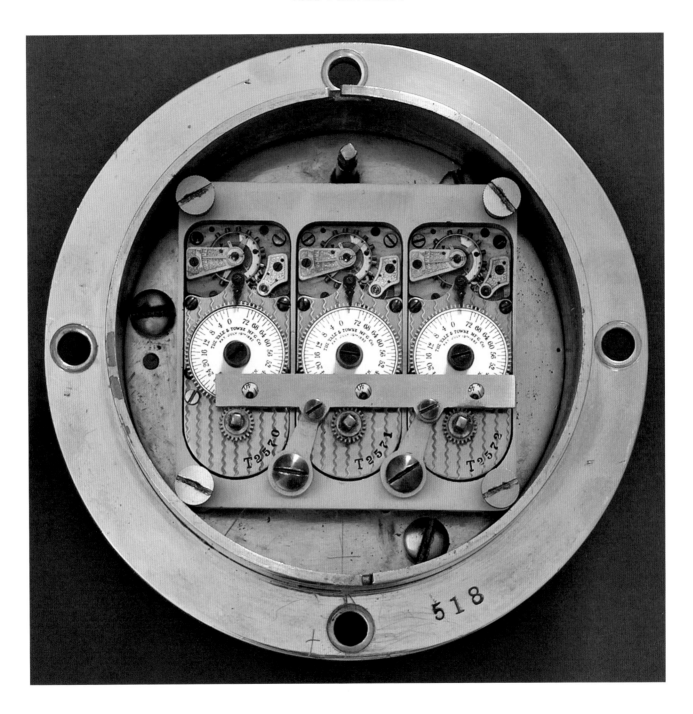

YALE & TOWNE L-MOVEMENT TIME LOCK

Through the late nineteenth and early twentieth centuries, the Victor Safe & Lock Co. of Cincinnati, Ohio, was one of the most prolific producers of freestanding safes in the United States. The cannonball safe was among the most successful designs among commercial safe owners and many safe makers introduced their own versions. The cannonball safe made by Victor Safe & Lock was one of the most popular, with its screw-door mechanism that allowed for one of the tightest-fitting doors without the need for excessive closing pressure. With the use of explosives to open safes becoming more prevalent in the 1890s, such a tight-fitting door was a major security feature. Another feature favored by banks and businesses with regular hours was the "solid door" design, in which the safe had a time lock but no key or combination lock.

Victor's solid screw-door cannonball safes commonly featured a Yale time lock with three seventy-two-hour L-movements made by Seth Thomas. Without any through-door connections for a wrench or bolt handle, the time lock itself included a light-powered automatic to retract the four door bolts. In keeping with the style of the time, Victor furnished an acid-etched mounting plate with an unusually beautiful art nouveau design. As many as one thousand of this time lock are thought to have been made, but fewer than five are known to survive today with an intact door-release mechanism.

Round-door bank chests were common and popular tools of small banks throughout the United States and the Victor Safe & Lock Co. was an important maker in the early twentieth century. This Yale time lock was designed for Victor Safe & Lock as a solid-door safe lock. Based on the Triple L model, it was mounted with a low-profile automatic that occupied the entire mounting space. The four bolts can be seen spaced evenly around the circumference of the case.

BANKERS DUSTPROOF 1906 TIME LOCK

With the time lock representing a major wholesale expense for any safe maker, many sought to have a time lock subsidiary, either by developing their own or by purchasing a going concern. By 1906 Victor Safe & Lock was offering its own time locks through its subsidiary, Bankers Dustproof,[16] and would continue until 1916. Some circumstantial evidence suggests that Bankers Dustproof was the result of Victor Safe & Lock's purchase of Consolidated Time Lock. Consolidated stopped production in 1906 and Bankers Dustproof began production in that same year; further, Bankers Dustproof movement designs are almost exactly the same as those of later Consolidated modular movements, the only major difference being the fixed or turning dial. Yet other factors suggest otherwise.

The patent underlying the early Bankers Dustproof model shown here[17] was filed in 1907 but languished in red tape for over two years, possibly because it openly admitted that it was a patent for the same outcome from "a less number of parts and simplified arrangement."[18] With so many important patents available to a company that took over Consolidated, one would have to wonder why Bankers Dustproof would sink two years of effort into securing such a dubious one.

This 1906 Bankers Dustproof time lock retains the original mounting plate from its Victor cannonball safe and controls a light automatic wound by the small arbor visible below the left-most movement. Early production included a glass door insert with later examples having a plain nickel-plated

BANKERS DUSTPROOF 1906 TIME LOCK

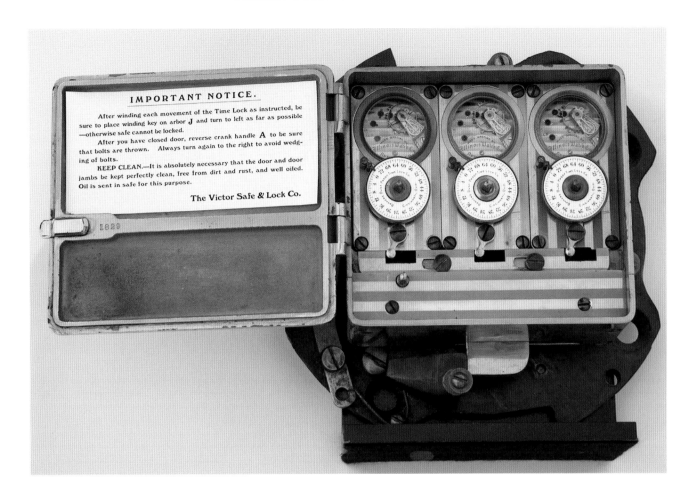

This Bankers Dustproof time lock was introduced around 1906 and featured its own internal automatic bolt motor, wound with the arbor below the first movement. Seen here still attached to the mounting plate and mechanism from a Victor solid-door model, this would have been the only lock on the safe. The decoration on the case and movements is clearly different from earlier periods, taking on the clean, art deco style of the time. Earlier examples of this model featured a glass door insert, although the solid door was likely an option for safes with coin traffic.

OPPOSITE AND ABOVE: *This Bankers Dustproof time lock from the early 1900s had the common glass door insert but also a side release that ran behind the movements. These side-release Bankers Dustproof time locks were generally intended for use in vaults with manual bolts in lieu of large automatics, a style increasingly popular after 1910. With the exit of E. Howard from the movement-making business in 1902, Bankers Dustproof relied on the Illinois Watch Co. for its movements.*

door and finally the acid-etched door seen here. Many examples of the 1906 Bankers Dustproof time lock have survived.

Other Bankers Dustproof models had four movements for use in vault doors. After the 1902 sale of E. Howard and the associated end to its time lock movement production, a major difficulty facing time lock makers seems to have been a consistent supply of movements. With the Illinois Watch Company committed to supply Bankers Dustproof, producing three- and four-movement models was not the financial risk it may have been for other makers. However, actual production of these models was always small and these models were never intended to operate the most powerful automatics. Instead, the Bankers Dustproof time locks used a bolt-dog design in place of the standard light automatic, even on models intended for use in vaults. Only one four-movement and fewer than ten three-movement Bankers Dustproof time locks for vault doors are known to remain.

BANKERS DUSTPROOF 1906 TIME LOCK

SARGENT & GREENLEAF'S CARY SAFE AUTOMATIC TIME LOCK

The Cary Safe Company of Rochester, New York, was a significant safe maker and, consequently, a significant reseller of time locks in its safes. Despite its strong relationship with Rochester cohabitant Sargent & Greenleaf, Cary used time locks from a number of makers. Made around 1908, this Sargent & Greenleaf time lock from a Cary Manganese Cannonball Bank Chest uses a Hall's Safe Company automatic bolt motor originally designed to work with a Consolidated time lock. As Hall's Safe & Lock had been the parent company of Consolidated Time Lock Company, this model is a curious hybrid, using parts made by direct competitors.

When this time lock was first introduced circa 1904 it was made with Consolidated time lock movements. But with the demise of Consolidated around 1907, these movements were no longer available. Cary's solution was to order a time lock from Sargent & Greenleaf that both fit the existing Cary safe design and could accommodate the bolt motor from the resurrected Hall Safe Co. The lock features three seventy-two-hour factory-modified Sargent Type H movements with special release levers operating the Hall bolt motor in a single bronze case. Sargent's manufacture of this lock is suggested by the surface jeweling common to Sargent locks over the entire case, except the bolt motor, which was supplied by Hall.

The original Consolidated version used three seventy-two-hour black-dial movements made by the South Bend Watch Co. Two examples of the Consolidated-made version of this time lock and one example of the Sargent-made version (shown here) are known to exist today.

SARGENT & GREENLEAF'S CARY SAFE AUTOMATIC TIME LOCK

An interesting hybrid, this assembly came from a Cary Safe Company round-door bank chest made around 1908. Cary Safe began offering this solid-door safe using a bolt motor made by Hall Safe Co. (not the then-defunct Hall's Safe & Lock) and a Consolidated time lock much the same as the time lock seen here. With Consolidated's demise around 1907, Cary looked to Rochester cohabitant Sargent & Greenleaf for a replacement. This time lock, made by Sargent, is essentially a copy of the Consolidated design and was used by Cary Safe until the firm could make the transition to using standard Sargent designs.

YALE & TOWNE Y-361 TIME LOCK

In 1906, the Ely Norris Safe Co. of Perth Amboy, New Jersey, introduced a new safe line, the Manard safe. Combining a number of time-tested security techniques, the Manard safe was a cannonball design with a time lock–only solid door that the user shut with a compression bar, minimizing space between the door and jamb, and included a "double compound" door design as well as manganese steel alloy construction, leading Ely Norris to bill the Manard as the "Strongest Safe in the World."

Whether this claim was true or just puffing for their advertisements, the Manard safe line was quite successful, and beginning in January 1909 Yale & Towne Mfg. Co. began supplying Ely Norris with an exclusive time lock design, the Yale Y-361 time lock. The Y-361's seventy-two-hour trioval-shaped Y-movements were made by Seth Thomas and controlled a light automatic bolt motor inside the case. Unlike Yale's Type B, C, D, and E time locks, each Y-movement has its own winding arbor accessed through an eyelet in the glass.

The Manard safe design continued in production into the 1920s, even after Ely Norris was taken over by the York Safe Co. of York, Pennsylvania. By the end of this production run, some three thousand Yale Y-361 time locks had been delivered[19] with four different dial designs that corresponded to the concurrent L-movement dials. The example shown here has the last of these four styles. A few hundred Yale Y-361 time locks survive today. Evidence exists for a concurrently made model denoted the Y-261, likely a less expensive model similar to the Y-361 that used only two Y-movements. Though there is no known photo, engraving, illustration, or surviving example of the Y-261, Yale's production ledgers as well as scattered instruction and maintenance documents make clear that the Y-261 was produced and sold, if only in limited numbers.

The Y-361 time lock from Yale was one of the last major design changes in the time lock industry. Using three pie-shaped time movements from Seth Thomas, the Y-361 included its own internal automatic in the remaining area of the case. Like other round-cased time locks, the Yale Y-361 and its suspected two-movement cousin, the Y-261, were used in solid-door safes with no other lock. Although the Y-361 is known to have been used only by the Ely Norris Safe Co. and its successor, the York Safe Co., production was substantial, eventually numbering over three thousand. The automatic is released by the round snubber disc at the center of the time lock.

SARGENT & GREENLEAF CLEOH TIME LOCK

Almost all Sargent time locks operate by blocking the boltwork of the safe or vault door, but the Sargent & Greenleaf Cleoh (a variation of Sargent's Model 4) operates on the lock only. The design of this Model 4B drew heavily on the Model 1 from 1874. Though the Model 1 offered both a time and combination lock in a single-case format, the Cleoh made the same lock-controlling time lock available for safe makers using a separate combination lock.

The Sargent Cleoh is shown here controlling a four-tumbler Sargent combination lock designed specifically for use with this time lock, as originally mounted in a National Safe Co. cannonball safe. Of the fewer than two hundred of these time locks made for the National Safe Co. beginning in 1905,[20] this example, which dates from about 1910, is the only complete unit known.

CLEOH TIME LOCK

The solid-door safe was popular, but it was truly useful only for businesses that wanted unfettered access to the safe's contents throughout the business day. Many users still preferred the security of a combination lock but, with small doors on the popular bank chests, mounting space was at a premium. One solution was Sargent & Greenleaf's Cleoh, a Model 4B time lock controlling a small-format combination lock. First seen in 1905, the Cleoh was one of the few major production designs from Sargent that used a linkage between the time and combination locks.

MOSLER SAFE CO. FOUR-MOVEMENT TIME LOCK

By the early twentieth century, banks in major metropolitan centers were constructing some of the largest vaults ever made. Among the few safe and vault companies able to install such a massive undertaking was the Mosler Safe Co. Mosler was a result of a series of consolidations of earlier companies, such as Mosler Safe & Lock Co., Mosler Bahmann, and possibly Bankers Dustproof Time Lock Co. Bankers Dustproof was the time lock subsidiary of the Victor Safe & Lock Company,[21] and although Victor Safe & Lock continued in business until approximately 1930, Bankers Dustproof disappeared as a brand in 1915. This coincided with the 1916 appearance of Mosler time locks[22] based on designs very similar to those of Bankers Dustproof, using the same seventy-two-hour 18-size pocket watch movements of the Illinois Watch Co. Changes in the design included a larger dial with a geared winding arbor, reduced decoration, modified dial numbering, and some rare 120-hour models.

This Mosler four-movement time lock was among its first models, introduced in 1916 for use in a main bank vault door. The cast-bronze case is gold-plated with surface texturing, placing this example within the first year of production. By 1917 the gold-plated, surface-textured case seems to have been replaced with a smooth, polished bronze case. The four seventy-two-hour movements are nickel-plated and use the Illinois Watch Co.'s 18-size Model #4 pocket watch movement. This model was among Mosler's most successful, continuing in production until well into the 1930s and offered in three- and two-movement versions as well. After 1932, Mosler switched to the American Waltham Watch Co.'s 16-size pocket watch movement, using an adapter ring to make this smaller movement compatible with the earlier case.[23] Production records for the 1916 Mosler four-movement time lock do not survive, but some hundreds are thought to have been made over its production life. Only one four-movement example of this first type is known today.

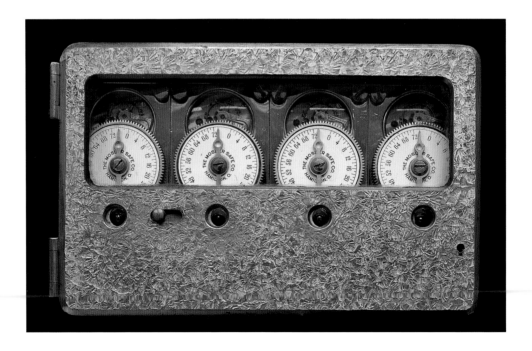

MOSLER FOUR-MOMENT

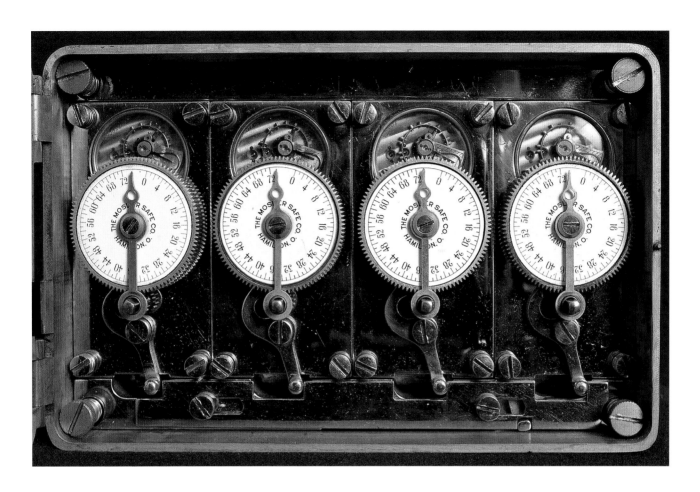

One of the earliest time locks offered by Mosler Safe Co. was this four-movement model that featured a gold-plated door with a textured surface, first seen in 1915. By this time the popularity of the automatic bolt motor for use in the largest vaults had begun to wane in favor of hand-operated boltworks. Consequently, this Mosler design was offered only with a side release, accepting a boltwork extension through the top right of the case, behind the movements.

CHAPTER 7

Alarm Timers

Among the most important advances in twentieth-century bank security was the electric alarm system. These alarms were sometimes connected to a central monitoring location by telephone lines, but they always included a timer to disable the alarm at a predetermined time, avoiding false alarms. Today, these alarm timers are often confused with time locks because they commonly include similar time movements. However, some familiarity with the most common formats of alarm timers will make them easily identifiable as the last part of the bank security equation to gain widespread acceptance.

NEW HAVEN CLOCK CO. ALARM TIMER

The adoption of mechanically timed electric alarm systems by banks at the end of the nineteenth century offered a new type of security and represented a new area of competition for both time lock makers and clock companies. Around 1890, the New Haven Clock Co. introduced a single-movement alarm timer, an example of which was donated to the Mossman Collection by Mosler. Although *Lure of the Lock* identifies this as "J. B. Young's Time Lock,"[1] this is not, in fact, a time lock. And despite its being marked "Young's Patent," this mechanism is not an example of either of the two time locks known to have been patented by Jackson B. Young in 1885.[2] Rather, it is likely that the New Haven Clock Co. owned the rights to Young's patents and included this claim on its alarm timers to bolster the company's credibility as a new entrant to the bank security business. Ultimately, however, a light, inexpensive brass case and a clock movement of modest quality yielded an alarm timer of questionable overall construction. The New Haven Clock Co. would not be a major maker of alarm timers, but it was among the vanguard trying to secure a place in this niche market.

Timers for the earliest electric alarm systems are often confused with time locks used in safes and vaults, but alarm timers such as this 1890 model from the New Haven Clock Co. are usually identifiable by their lighter insulated construction, single movement, and set of electric contacts. The electrical contact on this early design is visible extending from the top of the case. As the twenty-four-hour timer runs down, this contact will shift from the left to the right (as shown), breaking contact and turning off the alarm system.

NEW HAVEN ALARM TIMER

YALE & TOWNE ELECTRIC SWITCH TIME LOCK

In the detailed ledgers that the Yale & Towne Mfg. Co. maintained as part of its time lock interest pooling agreement with Sargent & Greenleaf, an interesting 1895 entry notes the production of an "electric switch time lock," numbered 1 through 10, made for the Chicago Edison Co., the electricity utility for Chicago.[3] We have no contemporary photos or illustrations of this Yale "electric switch," but the entire scope of Yale's timer business seems to be limited to this one model.

Among the Mossman Collection is a piece described in *Lure of the Lock* as a "circular time lock" by an "unknown maker."[4] However, this item is most likely an electrical timer, due to its single movement without any backup mechanism that operates a short lever, obscured by the timer's ornate rim. The beautiful, nickel-plated bronze case, enamel dial, and beveled glass cover indicate superior construction throughout, available from only a handful of companies. The single forty-eight-hour movement was made by E. Howard and is of a style known to have been used by Yale. And though there is no maker's attribution on the timer, Yale (unlike other manufacturers) is known to have produced a number of time lock lines with no attribution. The combination of a Yale-specific E. Howard movement, production quality worthy of a limited run or special order and on par with the best time locks, and the recently discovered record of Yale's sale to Chicago Edison all suggest that this may be an example of Yale's "electric switch time lock."

As with the timer that may have been Yale's Consolidated Edison electric switch, the maker of the alarm timer shown on page 332 has not yet been confidently identified. *Lure of the Lock* identifies this piece as an "Electric Time Lock,"[5] and while *Lure* correctly identifies Seth Thomas as the movement maker, it offers little else in the way of description.[6] However, a wired alarm contact is clearly visible at the movement top, showing this to be an alarm timer rather than an actual time lock.

Based on its basic, low-cost design with a plain brass case and thin-framed glass door, it may have been a working model and probably dates from about 1890 when makers were first working out alarm timer design. The unusual inclusion of two high-quality Seth Thomas movements is rather incongruous with alarm timers in general, but was even less likely in such a roughly finished model. One other timer employing a pair of Seth Thomas movements is known and these are currently thought to be early models made by the Seth Thomas Clock company itself, in an attempt to enter the burgeoning alarm timer market.

This high-quality mechanism was unidentified in 1928 when Lure of the Lock *was printed, but it is now thought to be an "electric switch" made by Yale for Chicago Edison in 1895. Although it is still unknown whether Chicago Edison used these timers to operate alarm systems or some other electrical mechanism, the style of construction is consistent with other alarm timers. The single forty-eight-hour E. Howard movement is set by the central winding arbor and controls a small release lever that is hidden by the broad, ornate rim.*

YALE ELECTRIC SWITCH

Working models of alarm timers are as rare as those of time locks. This example, from an unknown maker with its plain brass case and Seth Thomas clock movements, is one of the only known model alarm timers and was made early, likely around 1890. The two movements were an unusual extravagance for an alarm timer and may have made this design too expensive to go into actual production.

OPPOSITE AND OVERLEAF: *The earliest of the five alarm timers made by the Bankers Electric Protective Association was produced in 1896 and 1897, featuring one seventy-two hour E. Howard movement. The Bankers Electric design drew heavily on the well-understood time lock designs of the time, using a modular movement to actuate the contact much like the popular bottom release.*

BANKERS ELECTRIC PROTECTIVE ASSOCIATION ALARM TIMER

Among the most successful makers of bank alarm systems was Bankers Electric Protective Association, and between 1896 and 1901 Bankers Electric produced at least five distinct styles of alarm timer in both single- and double-movement versions.

The style thought to have been the earliest was made between 1896 and 1897, and was based on a seventy-two-hour movement made by E. Howard. This movement was substantially similar to the L-movement then in production by Howard for use in Yale's time locks but with a few notable differences: the main dial is marked with Bankers Electric's name, the front plate has the E. Howard attribution in engraved script rather than the Gothic type found on contemporary Yale movements, and the gold-plated parts of the escapement are unengraved. The case is nickel-plated bronze and a vertical armature extending through the case bottom is pushed left to right by the dial pin, breaking the circuit and disabling the alarm. Two examples of this single-movement alarm timer and one example of a double-movement version are known to survive today.

By 1898 the firm's name had been changed to the Bankers Electric Protective Company,[7] a change reflected on the dial of its second style of alarm timer, along with other changes: the attribution to "E. Howard Boston" on the movement

was now in Gothic type, the gold-plated escapements now featured elaborate engraving, the movement's front-plate damascening became wavy rather than angular, straight-line damascening now appeared on the case door, and the throw mechanism now incorporated a nickel-plated hinged lever. This style is thought to have been the subject of two orders for twenty-five movements with an option for seventy-five more, although it is not known whether this option was ever exercised.[8]

At least one other type of Bankers Electric alarm timer is known to have been made. One eighty-hour design exists, much in the Consolidated Time Lock Co.'s style, with a laterally rectangular front plate and a platform escapement. Its solid white dial marked "The Bankers Electric Protective Company"

BANKERS ELECTRIC ALARM TIMER

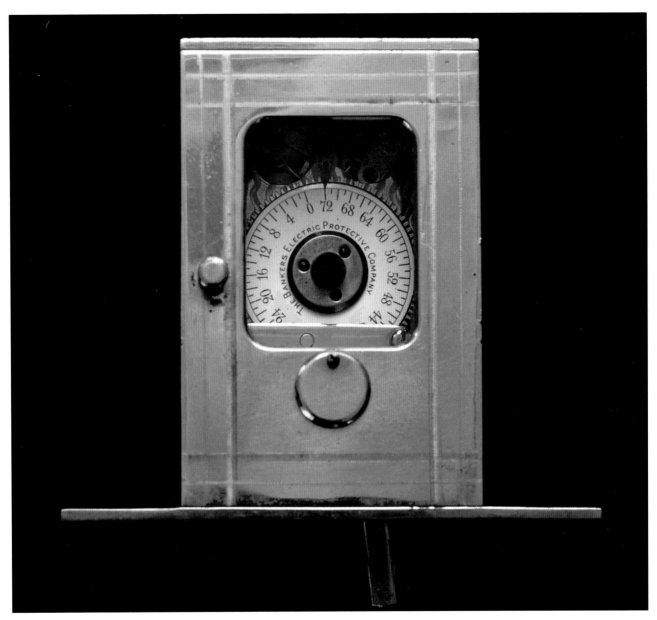

ABOVE AND OVERLEAF: *The second style of timer from Bankers Electric introduced a simpler, less expensive actuating arm across the movement dial. This example was made after 1898, as shown by the name of the maker having been changed on the dial from "Association" to "Company." As on the first model, the case includes an extended base that gave the user a mounting plate designed to be screwed onto the alarm's wiring case. Such a case would have been made of wood or another insulating material, making internal insulation unnecessary.*

implies that it was made after the 1898 name change, and its longer eighty-hour power reserve suggests that it was introduced after the company's second seventy-two-hour model. Another style of movement appears in an undated advertisement for the Bankers Electric Protective Company. The "automatic electric burglar alarm" is described as having a fifty-six-hour E. Howard–made movement, but surviving

Howard production records do not seem to include this unusual movement. Although the company name firmly places the flyer post-1898, the image and description may be of an earlier or possibly experimental model. With only a single photo and no production record, we cannot be confident that this model was more than an advertising sample. One example of a fifth, substantially similar type is part of the Harry Miller Collection.

AMERICAN BANK PROTECTION CO. ALARM TIMER

Another major bank alarm system maker at the end of the nineteenth century was the American Bank Protection Co., which produced two types of alarm timers. The earlier model was designed around the standard seventy-two-hour E. Howard L-size movement that was then in use in Yale's time locks. However, with the takeover of Howard by the Keystone Watch Case Co., L-movement production came to a total halt in 1902,[9] forcing American Bank Protection to go elsewhere for timer movements.

Most companies affected by E. Howard's exit from the movement market turned to Seth Thomas, and American Bank Protection was no exception. The company's post-1902 alarm timers employed a single-escapement Seth Thomas movement with a twenty-four-hour face. A clock speed adjustment

In 1902, E. Howard was taken over by the Keystone Watch Case Co., ending its production of movements for time locks and alarm timers. Most makers turned to Seth Thomas, and though Seth Thomas did introduce movements similar in design to the modular E. Howard models, American Bank Protection chose to adopt the standard clock movement. Seen here in a post-1902 alarm timer, the clock movement uses a twenty-four-hour face and an internal armature to connect and disconnect the three contacts on the case top.

AMERICAN BANK PROTECTION CO. ALARM TIMER

is visible at the top of the face. Notably, the electric mechanism of this later American Bank Protection timer is based closely on an earlier design originally patented by Edwin Holmes in 1867[10] and reissued in 1880.[11] Rather than suggesting a connection between Holmes and American Bank Protection, this similarity was more likely due to the recent expiration of the seventeen-year period of patent protection on Holmes's design. This style of timer was typically mounted in pairs in an elaborate wooden case, which also housed the direct current batteries that powered the alarm system.

AMERICAN BANK PROTECTION CO. ALARM TIMER

OPPOSITE: *The alarm timers made by the American Bank Protection Co., another major alarm company in the early twentieth century, were quite different in appearance from those of Bankers Electric. The Bankers Electric models were designed after the time locks that they grew from, but American Bank timers used spacious wooden cases with their own internal wiring. Rather than rely on the alarm box for insulation, the American Bank Protection timer is itself made of insulating materials, and features the electric contacts on the top, ready for connection to the alarm system.*

ABOVE: *One of American Bank Protection's design innovations was a new actuator shape. Based on the same single seventy-two-hour E. Howard movement as the Bankers Electric model, with a pin that slowly turned with the dial (visible at the dial bottom), the American Bank Protection design had an actuator arm with a curved extension that rode up on the pin, lifting the arm and disarming the alarm.*

DIEBOLD ALARM TIMER

After the growth of Diebold's market position in bank vault installation, bank locks, and time locks at the end of the 1890s, Diebold also sought to make significant inroads into the bank alarm system market. Production of Diebold's alarm timers most likely did not begin until after the 1902 takeover of E. Howard by Keystone; hence, all Diebold timers used movements from Seth Thomas.

Most bank security companies in the early 1900s were safe/lock/time lock makers or electric alarm system/alarm timer makers. Unlike these contemporaries, Diebold quickly became a major maker of a full range of bank security systems, including large vaults, high-quality time locks, and electric alarm systems. As was the case with its three- and four-movement time locks (which were the same basic design and used the same movements), Diebold adopted the same model of seventy-two-hour Seth Thomas modular movement that was found in the company's concurrently produced time locks. The alarm timer shown here was one of Diebold's earliest entries into the alarm timer market.

By the 1910s Diebold had become a full-service bank security company, offering a complete line of the highest-security safes, vaults, locks, and alarm systems. Unlike American Bank Protection Co., which adopted a Seth Thomas clock movement after 1902, Diebold used the Seth Thomas modular movement, seen here. With its small Bakelite case and curved actuating armature, the Diebold alarm timer was not particularly innovative, but rather represented the apex of alarm timer design prior to modern electronics.

DIEBOLD ALARM TIMER

EPILOGUE

The sun began to set on the golden age of American bank lock development with the beginning of World War I. With the industry having passed out of the hands of the first generation of bank lock innovators, the focus of manufacturers shifted from innovation to commerce and the major companies settled into the comfortable business of making, installing, and servicing their lock designs. With the United States' entry into the First World War in 1917, demands for iron, steel, and bronze rose sharply, leading to the wholesale scrapping of safes, often with their locks in place. Although many of these early scrapped safes were later replaced, the models available during the 1920s were little changed from those introduced ten years earlier.

The end of the 1920s brought the Great Depression, an economic catastrophe felt in every industry in the United States, none more sharply than the banking industry. New bank openings and new vault installations came to such a sharp halt that Sargent & Greenleaf, the flagship time lock maker, did not record a single domestic sale of a new time lock between 1929 and 1940. The World War II years were another period of war footing with its attendant demands for metals, leaving little for safes and locks. By the time the United States returned to an extended period of peaceful commerce, introduction of low-cost materials such as plastics, the maturation of the Federal Reserve System, and shifts in banking regulations all combined to make safes, vaults, and the locks that secured them far less important to the survival and success of banks. For a short period after the war, time lock movements were still made in the United States until production shifted to Switzerland and elsewhere beginning in the 1950s.

Contrary to the glamorized movie version of safecracking, there are really no known examples of safecrackers forcing their way through a high-security vault door. Modern vaults are simply too difficult to penetrate for all but the most sophisticated technicians with tools, time, and no alarms to worry about. There is, however, no defense against poor design, as shown by the June 2003 burglary of the Atlantic Bank of New York on Northern Boulevard in Flushing, New York. The bank's safe deposit vault boasted a formidable door and sound and motion sensors, yet its ceiling was a meager half-inch thick. Working over a weekend when the sensors were disabled, the safecrackers cut a hole into the top of the vault from the floor above and made off with significant amounts of cash. Though such instances are rarities, they serve as a reminder that bank security is an evolving process rather than a onetime project.

This evolution saw the introduction of the first high-security electronic safe and time locks during the 1980s and '90s. While reliability was initially lower than the best mechanical systems, major progress has been made and the possibility of combining these electrical systems with computerized control and biometric identification makes them the standard to come. One example of a security system that incorporates this type of multilevel electronic control is the Navigator system for ATM machines from La Gard, Inc. The authorized user is equipped with a smart phone that connects with both the lock and a central server. A live administrator authenticates the phone with a user code and transmits a one-time combination to the user. The user can unlock the device with this combination for a short period, during which the time, date, and duration of entry are recorded. With 128-bit keyless electronic encryption combined with the short window of key validity, La Gard's Navigator is immune to all known hacking methods. Such is the future of bank vault technology, and though there will always be those who work to plunder what has been amassed, for the present the bankers seem to be one step ahead.

Diebold factory floor circa 1918. With their steel foundries and heavy equipment, safe makers were well positioned to aid the war effort after some retooling. Here, a Diebold factory has been modified to produce light tanks during World War I.

NOTES

CHAPTER 1: The American Bank Lock Industry

1. Alfred Hopkins, *The Lure of the Lock* (New York: General Society of Mechanics and Tradesmen, 1928), pp. 39–40.
2. E. Monk, *Keys—Their History and Collection* (Buckinghamshire: Shire Publications, 1999), p. 27.
3. Hopkins, *Lure of the Lock*, p. 50. *See also* UK Pat. No. 4,219, Jeremiah Chubb, February 1818.
4. A complete description of Hobbs's triumph and technique can be found in *Lure of the Lock*, from p. 48.
5. *The Times*, London, September 4, 1851.
6. Federal taxes on private banknotes of 10 percent began in 1865 as a way of dealing with the Civil War debt. Later changes such as deposit insurance, consistent inflation, and the bullion-ownership prohibition between 1933 and 1975 all raised the costs associated with holding gold and redeemable money. Ultimately gold and bank-issued notes were rendered impractical as a monetary standard.
7. William Greer, *A History of Alarm Security*, 2nd edition (Bethesda, Md.: National Burglar & Fire Alarm Association, 1991), p. 2.
8. Hopkins, *Lure of the Lock*, p. 86.
9. Ibid.
10. Ibid., p. 123.
11. These "recovered patents" are available on microfilm at the Science, Business and Industry Library of the New York Public Library. These records are not an exhaustive reproduction of pre-1936 American patents, since the British patent officials retained copies only of American patents that were of specific interest, such as those that were part of British patent court cases.
12. Unnumbered patent, awarded January 20, 1830, "Permutation Lock."
13. Thomas Hennessey, *Locks and Lockmakers of America* (Park Ridge, Ill.: Locksmith Publishing Corp., 1997).
14. Ibid.
15. *Rochester Union & Advertiser*, July 2, 1892.
16. Ibid.
17. Hopkins, *Lure of the Lock*, p. 191.
18. Ibid.
19. Ibid., p. 192.
20. Ibid., pp. 194–95.
21. UK Pat. No. 6,105, 1831.
22. *Interference*; application for reissue of John Burge (filed November 17, 1876) and application for reissue of James Sargent (filed April 5, 1876); subject "Time Lock."
23. *See, e.g.*, the Sherman Antitrust Act of 1938 (15 U.S.C. §§ 1, 2).
24. *Contract Respecting Time Locks*; entered into October 16, 1877, by Sargent & Greenleaf and Yale Lock Mfg. Co.
25. Ibid., §§ I–V.
26. Section X, *viz.*, Sargent No. 1, 2 and 3—$300, $250 and $250 wholesale, $500, $400, and $400 retail, respectively; Yale Pin Dial with and without Sunday Attachment—$275 and $250 wholesale, $450 and $400 retail, respectively.
27. Section XII, *viz.*, Sargent controlled ME, NH, MA, RI, NY, NJ, MO, IA, OH, IN, MI, MN, the District of Columbia, PA west of 78°W long., IL south of 40°N lat., and a 50 percent interest in the city of Chicago; Yale controlled CT, DE, MD, VA, WV, NC, SC, GA, FL, AL, MS, LA, TX, WY, KS, NE, AR, KY, TN, CA, NV, OR, ID, UT, AZ, CO, MT, NM, "Dakota," the "Washington Territory," the balance of PA and IL, and the remaining 50 percent interest in the city of Chicago.

28. Section VII, *viz.* Pillard, $250; Holmes, $250; Hall, $200; Herring, $200; Lillie, $200; and "others at proportionate rates."
29. Section XI. This profit-splitting arrangement required that the records be provided and settled monthly and made under oath. Much of Yale's production records survive today, providing a wonderful source of original material, thanks to their preservation by the Lock Museum of America in Terryville, Connecticut.
30. *See Sargent & Greenleaf vs. The Yale Lock Manuf'g Company*, Award of Arbitrators on Time Lock Patents, March 2, 1878.
31. *Yale Lock Mfg. Co. v. Berkshire Nat'l Bank*, 135 U.S. 342, 1890.
32. *Yale Lock Mfg. Co. v. Norwich Nat'l Bank*, 19 Blatch. 123, 1881; *Yale Lock Mfg. Co. v. New Haven Savings Bank*, 6 Fed. Rep. 377, 1881.
33. M. Joblin & Co., *Cincinnati Past and Present: or Its Industrial History, as Exhibited in the Life-Labors of its Leading Men* (Cincinnati, Ohio: Elm Street Printing Co., 1872), p. 185.
34. Ibid., pp. 186–87.
35. *See Hall v. MacNeale Urban*, Dist. Ohio, 1875, deposition of Joseph Hall.
36. Joblin & Co., *Cincinnati Past and Present*, p. 187.
37. Ibid., p. 186.
38. Ibid.
39. *Hall v. MacNeale & Urban Co.*, 1875, deposition of Milton Dalton.
40. Ibid.
41. Joblin & Co., *Cincinnati Past and Present*, p. 186.
42. *Yale Lock Mf'g Co. v. Berkshire Nat'l Bank*, Dist. Mass., 1890.
43. Fallon, Feltzer, Schapiro, *Hart and Wechsler's The Federal Courts and the Federal System*, 4th edition (Westbury, N.Y.: The Foundation Press, 1996), pp. 35–38.
44. *See, e.g.,* Moore and Levi, "Federal Intervention," 45 *Yale L. J.* 565 (1935–36); *see also* Oliver Shiras, *Equity Practice in the United States Circuit Courts* (Chicago: Callaghan and Company, 1898).
45. The trial court for many patent infringement actions was a federal circuit court, a notable difference from today's federal system in which the district court is the trial-level venue and the circuit courts handle only appeals.
46. 17 Fed. Rep. 531.
47. 26 Fed. Rep. 104.
48. *Yale*, 135 U.S., at 345.
49. Ibid., pp. 378–79.
50. Ibid., p. 379.
51. *See, e.g.,* Dan Graffeo, "Consolidated Time Locks," *Safe and Vault Technology*, November 1990, pp. 19–23.
52. *See* Public Notice, Hall's Safe Co., 1906.
53. Greer, *A History of Alarm Security*.
54. Reissue Nos. 8,753 and 8,754; June 17, 1879.
55. *Holmes Burglar Alarm Telegraph Co. v. Catskill Nat'l Bank*, Cir. Ct. S.D.N.Y., Case No. C-1259, filed 1879.
56. *Holmes Burglar Alarm Telegraph Co. v. Connecticut Mut. Life Ins. Co.*, Cir. Ct D.C.T., filed July 11, 1879.
57. Ibid.
58. *Scientific American*, November 17, 1894.
59. *Scientific American*, April 21, 1906.

CHAPTER 2: The Early Prominence of Key Locks: 1834–1856

1. Hopkins, *Lure of the Lock*, p. 108.
2. *Mechanics' Magazine and Register of Inventions and Improvements*, vol. III, no. 5; May 31, 1834, p. 32.
3. Patent of January 11, 1836, of S. Andrews. The patent papers are lost and the number is not known.
4. Lynn Collins interview, November 20, 2004.
5. Ibid.
6. Ibid.
7. *The Flying Jerseyman*; advertisement cutting from Newark newspaper of June 30, 1842, proceedings of the New Jersey Historical Society, p. 165.
8. Although the mechanism of the Snail Wheel Lock uses disc-shaped tumblers, it is not related to the "disc tumbler" lock common today. The modern disc tumbler lock is a less expensive, less secure relative of the pin tumbler lock, using a spring to force projections from each disc to block the rotation of a cylinder. For a useful discussion of the modern disc tumbler mechanism, see Max Alth, *All About Locks and Locksmithing* (New York: Hawthorn Books, 1972), ch. 5.
9. Edmund Beckett Denison, M.A., Q.C., *Locks and Clocks from the "Encyclopædia Britannica,"* 2nd edition (Edinburgh: Adam and Charles Black, 1857), p. 205.
10. Ibid., pp. 206–7.

11. Hopkins, *Lure of the Lock*, p. 62.
12. Pat. No. 3,630, June 13, 1844, L. Yale, "Door Lock."
13. Pat. No. 6,111; February 13, 1849, L. Yale, "Lock and Latch."
14. There has been some suggestion that Yale's design was the first instance of a lock controlling another lock, rather than double custody relying on two locks independently securing a door. See, e.g., Hopkins, *Lure of the Lock*, p. 61, "Here we have the germ of a great idea, having one piece of machinery guard the actions of another." However, it is now generally accepted that Yale's design was just the first to gain wide recognition.
15. Others included marble dust, alum, wood, wood ash, paper pulp, water, water cans or tubes opened by heat to allow for boiling, sheet mica, lime-free "Rosendale" cement (which resisted swelling and bursting), water lime, and air-slacked lime. See Milton Dalton, *History of Fire & Burglar Proof Safes, Bank Locks and Vaults in America and Europe—Useful Information for Bankers, Business Men and Safe Manufacturers* (Cincinnati, Ohio: Franklin Type Foundry, 1874).
16. Dalton, *History of Fire & Burglar Proof Safes*, p. 35.
17. Ibid.
18. Notable exceptions are listed in Dalton's *History*, p. 73 ("Combination Fire and Burglar-Proof Safe" offered by J. McB. Davidson of Exchange Street, Albany, NY, 1872), and p. 88 ("Lillie Fire and Burglar-Proof Safe" offered by Lillie through Fowler in Chicago until at least 1867).
19. *Scientific American*; June 11, 1853, p. 307.
20. Thomas B. Leman, *The Historical Significance of the Salamander and its Relationship with Asbestos and the Asbestos Worker*, unpublished paper, April 15, 2003.
21. Pat. No. 3,747, September 17, 1844, R. Newell, "Lock."
22. Hopkins, *Lure of the Lock*, p. 56 (legible with some effort at the bottom of the reproduced advertising schematic).
23. Ibid., pp. 114–15.
24. *Boston Post,* April 27, 1859, p. 1.
25. Pat. No. 15,031, June 3, 1856, L. Yale, "Lock."
26. Pat. No. 6,878, April 3, 1849, D. M. Smith, "Bank Lock."
27. "Judges' Report on Bank Locks," 24th Fair of the American Institute, New York, 1851.
28. See, e.g., the patents of J. R. and H. C. Campbell (1835), E. Finney (1839), or D. W. Maples (1844).
29. Silas C. Herring, *Interesting and Important Information Respecting the Preservation of Books, Papers, Money, Jewelry, &c., from the Ravages of Fire [/] The Unequalled Security Afforded by Herring's Fire-Proof Safes* (New York, 1855), p. 131.
30. Hopkins, *Lure of the Lock*, p. 158.
31. Herring, *Interesting and Important Information*, p. 3.
32. Hopkins, *Lure of the Lock*, p. 122.
33. Dalton, *History of Fire & Burglar Proof Safes*, p. 59.
34. Ibid.
35. Hopkins, *Lure of the Lock*, p. 166.
36. Dalton, *History of Fire & Burglar Proof Safes*, p. 109.
37. *Sargent v. Hall Safe & Lock Co.*, Case No. 178, Term 1884, Appellees' Exhibit, Stipulated Appendix to Exhibits Certified to Supreme Court, pp. 581–83 ("Judges' Report on Bank Locks").
38. *Yale Lock Mfg. Co. v. New Haven Savings Bank*, 6 Fed. Rep. 377, 1881.
39. *Yale Lock Mfg. Co. v. Norwich Nat'l Bank*, 19 Blatch. 123, 1881. (Francis B. Pye made and showed full-sized version in 1850 or 1851.)
40. Pat. No. 4,406, March 7, 1846, F. B. Pye, "Lock."
41. Dalton, *History of Fire & Burglar Proof Safes,* p. 112.
42. Pat. No. 6105, July 2, 1831, "Apparatus to be Applied to Locks and other Fastenings."
43. Pat. No. 5321, October 9, 1847, "Vault Lock."
44. Hopkins, *Lure of the Lock*, p. 128.
45. These included false-rivet buttons and movable back plates, with some locks requiring many steps to open. For an excellent overview, see Robert W. Dix, "The Locks of H. C. Jones", *Journal of Lock Collecting*, vol. 33, no. 3, May 2002, p. 3.
46. Ibid., p. 9.
47. Pat. No. 6,252, April 3, 1849, H. Ritchey, "Lock."
48. Pat. No. 4,011; April 26, 1845, H. C. Jones, "Bank Lock."
49. Herring, *Interesting and Important Information*, p. 131.
50. *Lure of the Lock* identifies him as "Betterly" (p. 119); "Betteley" is proper. See, e.g., Betteley Deposition, *Hall v. MacNeale Urban*.
51. Hopkins, *Lure of the Lock*, p. 119. Despite Hopkins's misspelling of the maker's name, nothing in *Lure of the Lock* contradicts this date of manufacture.
52. Pat. No. 8,918, April 6, 1852, Betteley, "Door Lock."
53. Pat. No. 8,071, May 6, 1851, L. Yale, Jr., "Lock and Key."

54. Pat. No. 32,331, June 4, 1861, L. Yale Jr., "Lock."
55. Hopkins, *Lure of the Lock*, p. 107.
56. Dalton, *History of Fire & Burglar Proof Safes*, p. 56.
57. Pat. No. 10,144, October 18, 1853, L. Yale, "Lock."
58. Pat. No. 11,158, June 27, 1854, W. Hall, "Door Lock."
59. Hopkins, *Lure of the Lock*, p. 111.
60. Ibid., p. 125.
61. Collins interview, November 20, 2004
62. Ibid.
63. *Lure of the Lock* lists Brennan's key lock as patented on November 7, 1885, but this was likely a typographical error; November 7, 1854, is proper. See Pat. No. 11,885; November 7, 1854; "Lock," J. B. Brennan.
64. Hopkins, *Lure of the Lock*, pp. 113–14.
65. Pat. No. 67,927, August 20, 1867, T. J. Sullivan, "Permutation Lock for Doors."
66. Pat. No. 12,932, May 22, 1855, L. Yale, "Lock."
67. Hopkins, *Lure of the Lock*, p. 87.
68. Ibid.
69. Pat. No. 3,630, June 13, 1844, L. Yale, "Door Lock."
70. Pat. No. 4,270, November 12, 1845, "Perm[utation] Lock."
71. Pat. No. 15,239, July 1, 1856, "Permutation Lock."
72. Ibid., p. 2.
73. For an enjoyable, anecdotal account and images of this second version of Isham's Permutation lock, see *Journal of Lock Collecting*, Jan./Feb. 2004, p. 9.
74. Pat. No. 17,740, July 7, 1857, "Permutation Lock."
75. Dalton, *History of Fire & Burglar Proof Safes*, p. 43.
76. Ibid. The actual patent for this Krenkel design has not been identified.
77. Hopkins, *Lure of the Lock*, pp. 109, 112.
78. Ibid., p. 109.

CHAPTER 3: The Combination Lock Comes of Age: 1857–1871
1. Pats. No. 18,228, September 18, 1857; 21,865, October 5, 1858; 23,222, March 1, 1859, Henry W. Covert.
2. Hopkins, *Lure of the Lock*, p. 150.
3. Pat. No. 17,150, April 28, 1857, Holbrook & Fish, "Chronometric Lock."
4. *Milford Journal*, February 14, 1857.
5. Incorporating Pat. No. 19,927, April 13, 1858, A. Holbrook, "Chronometric Lock."
6. Deposition of Amos Holbrook ("Holbrook Deposition"), November 6, 1878, *James Sargent v. Hall Safe & Lock Co.*, before U.S. Cir. Ct., Southern Dist. of Ohio, as part of Exhibits to U.S. Sup. Ct., p. 483.
7. Ibid., p. 484.
8. Sargent & Greenleaf, Inc., *Bank Locks, Catalog No. 21*, Rochester, N.Y., 1927, p. 8.
9. Hopkins, *Lure of the Lock*, p. 180 (reproduced in toto from Sargent's *Catalog No. 21*).
10. *See, e.g.*, David Christianson, "Time Locks: Their History from Beginning to End," *NWACC Bulletin*, vol. 46/6, no. 353, December 2004, p. 741 (citing *Lure of the Lock*).
11. Illustrated Holbrook Circular, *Sargent v. Hall Safe & Lock Co.*, No. 178, Term 1884, Appellees' Exhibit, in Stipulated Appendix to Exhibits Certified to Supreme Court, pp. 600–5.
12. Deposition of Alvin Underwood, undated, *James Sargent v. Hall Safe & Lock Co.*, before U.S. Cir. Ct., Southern Dist. of Ohio, as part of Exhibits to U.S. Sup. Ct., p. 515.
13. Deposition of Samuel Hayward ("Hayward Deposition"), November 6, 1878, *James Sargent v. Hall Safe & Lock Co.*, before U.S. Cir. Ct., Southern Dist. of Ohio, as part of Exhibits to U.S. Sup. Ct., p. 521.
14. Holbrook Deposition at 491–92.
15. Ibid., p. 483.
16. Hayward Deposition at 526.
17. Holbrook Deposition at 483–84.
18. See *Sargent v. Hall Safe & Lock*, Appendix to Exhibits, pp. 891–92 (deposition of Joseph L. Hall, September 5, 1879).
19. *Milford Journal*, June 9, 1875.
20. Hayward Deposition at 526.
21. Pat. No. 21,346, August 31, 1858, Lord, "Gear Lock."
22. Pat. No. 21,689, October 5, 1858, L. H. Miller, "Permutation Lock."
23. Pat. No. 28,162, May 8, 1860, L. Derby, "Combination Lock." (*N.B.* On the drawings page of this patent specification, two figures may have been numbered "*Fig. 5.*" The upper figure should probably be "*Fig. 3.*")
24. Ibid., pp. 2–3.
25. Hopkins, *Lure of the Lock*, p. 154.
26. Pat. No. 28,710, June 12, 1860, L. Yale, Jr., "Lock."

27. Pat. No. 32,331, May 14, 1861, L. Yale, Jr., "Lock." Reissued Pat. No. 1,469, April 28, 1863.
28. Hopkins, *Lure of the Lock*, p. 126 (catalogue reproduction).
29. Ibid.
30. Pat. No. 48,475, June 27, 1865, L. Yale, Jr., "Lock."
31. Pat. No. 43,457, July 5, 1864, N. MacNeale, "Bank Lock."
32. Hopkins, *Lure of the Lock*, p. 126 (catalogue reproduction).
33. Ibid.
34. Pat. No. 47,575, May 1, 1865, Sargent & Covert, "Lock." Pat. No. 51,973, January 9, 1866, Sargent & Covert, "Lock."
35. Hopkins, *Lure of the Lock*, p. 137.
36. Catalogue of Sargent & Greenleaf, 1867. *See also*, Hopkins, *Lure of the Lock*, p. 137.
37. Pat. No. 20,658, June 22, 1858, S. Perry, "Permutation Lock."
38. "New Britain Bank Lock Company," *New Britain Record*, July 6, 1866.
39. *Directions for Using the Key Register Bank and Safe Lock*, New Britain Bank Lock Company; ca. 1860.
40. Pat. No. 17,293, May 12, 1857, "Lock." Pat. No. 20,658, June 22, 1858, S. Perry, "Permutation Lock."
41. "New Britain Bank Lock Company."
42. Pat. No. 108,134, October 11, 1870, Henry Gross, "Permutation Lock."
43. Harry C. Miller, *Harry C. Miller Lock Collection, Collection Number 200* (Nicholasville, Ky.: Lockmasters, 1976), p. 1.
44. Ibid., p. 2.
45. Ibid.
46. *Sargent v. Hall Safe & Lock Co.*, Appendix to Exhibits, pp. 891–92 (deposition of Joseph L. Hall, September 5, 1879).
47. *Hall v. MacNeale & Urban Co.*, deposition of Jacob Weimar.
48. Hopkins, *Lure of the Lock*, p. 130.
49. Ibid., p. 159, incorrectly identifying maker as "Krinzle"; "Kienzle" is correct. *See* Dalton, *History of Fire & Burglar Proof Safes*, p. 95 (February 7, 1873, sworn statement of Carl Diebold, "of the firm Diebold & Kienzle").

CHAPTER 4: The Rise of the Time Lock: 1872–1888

1. *Interference*; Burge and Sargent (1876), statement of Laporte Hubbell, September 9, 1875, pp. 79–83.
2. Ibid.
3. UK Pat. No. 6,105, 1831.
4. Canadian Pat. No. 2,190, 1873.
5. *Interference*; Burge and Sargent (1876), statement of John Burge, August 13, 1875, pp. 8–16.
6. Ibid., pp. 21–23.
7. Pat. No. 194,506, August 21, 1877, J. Burge, "Time-Locks."
8. For a more complete discussion of Schroder's locks, see the American Lock Collectors Association's *Journal of Lock Collecting*. Robert W. Dix, "Schroder Lock Co.," vol. 33, no. 4, July/August 2002, pp. 3–9
9. Lynn Collins interview, March 7, 2005.
10. Ibid.
11. Ibid.
12. Ibid.
13. *Interference*; Burge and Sargent (1876), Exhibit L, December 9, 1875, p. 41.
14. Ibid.
15. Ibid.
16. For all Sargent Model 2 time locks, our notation will include the type in brackets. Sargent did not specifically denote any style differences in the Model 2, despite its significant evolution over its fifty-five-year life.
17. Sargent & Greenleaf Model 2 instruction sheet.
18. Pat. No. 161,283, March 23, 1875, J. Sargent, "Bolt for Safe, Vault and other Doors."
19. Hopkins, *Lure of the Lock*, p. 180.
20. Sargent & Greenleaf, Inc., *Catalog No. 21*, 1927, p. 8.
21. *See* Hopkins, *Lure of the Lock*, p. 180. *See also*, Sargent & Greenleaf, *Catalog No. 21*, p. 8.
22. *See* Pat. Reissue 7,835 (August 7, 1877); Pat. No. 195,539 (September 25, 1877); Pat. Reissue 7,947 (November 13, 1877).
23. Sargent & Greenleaf, *Catalog No. 21*.
24. Ibid.
25. Sargent & Greenleaf, *Sargent & Greenleaf Sale Records 1920–1939*.
26. Pat. No. 168,062, September 9, 1875, Emery Stockwell, "Time-Lock."
27. *Interference*; Burge and Sargent (1876), deposition of Emery Stockwell, November 15, 1876, pp. 92–93.

28. Howard Ledgers.
29. Ibid., vol. 8, p. 101.
30. Yale advertising circular, 1875; Yale catalogue, 1896.
31. Howard Ledgers.
32. Ibid.
33. Ibid., vol. 9, pp. 1, 56.
34. *Yale Lock Mf'g Co. v. Berkshire Nat'l Bank*, Dist. Mass., 1890, deposition of Joseph Hall.
35. *See Hall v. MacNeale Urban*, Dist. Ohio, 1875, deposition of Henry Gross.
36. Pat. No. 173,121, February 8, 1876, H. Gross, "Time Attachment for Locks."
37. The portions of the production records of the E. Howard Watch & Clock Co. that survive have helped shed light on the activities of many time lock makers, since at one time or another most found E. Howard to offer the best balance of reliability and cost. Many insights into the time lock industry stem from our analysis of these records, which were preserved by and reviewed with the gracious permission of Dana Blackwell. These ledgers have been donated to the Smithsonian Institution and we refer to them collectively in our references as Howard Ledgers.
38. Howard Ledgers, vol. 2, p. 53.
39. Ibid., vol. 9, p. 81.
40. *Viz.*, the shape and mounting of the balance wheel hairspring stud arm, the regulator design, and the mounting plate engine turning are all of a style unknown among E. Howard movements.
41. Ibid., vol. 4, p. 153.
42. *Yale v. E. Howard Watch Co.*, probably District of Massachusetts, ca. 1883–4. Sadly, the record of this important lawsuit is incomplete, and the limited documents that exist at the Massachusetts federal archives cannot even give an accurate citation. No doubt, the testimony of E. Howard's executives would have been a trove of information.
43. *See Interference*; deposition of Emery Stockwell, November 15, 1876, p. 90.
44. Ibid., pp. 92–93.
45. Ibid.
46. Pat. No. 174,996, March 21, 1876, Charles O. Yale, "Time Lock."
47. Pat. No. 210,070, November 19, 1878, Jacob Weimar, "Time-Lock."
48. From a Herring sales flyer delivered with a double-dial model Dexter, dated to 1876 by handwritten notation.
49. *Sargent & Greenleaf v. National Mohawk Valley Bank*, U.S. Cir. Ct. (N.D.N.Y.), testimony of Henry L. Brevoort (stating that trial model was same as one purchased from Herring & Co. and used by bank).
50. *Yale Lock Mfg. Co. v. Berkshire Nat'l Bank*, 135 U.S. 342, 1890.
51. *Hall v. MacNeale & Urban Co.*, 1875, affidavit of E. J. Woolley.
52. *Lure of the Lock* identifies the bracket as a "pedestal" and suggests that some parts are missing. However, a careful comparison with the patent specification has shown that it is in fact complete, including liquid inside the timing chamber.
53. Pat. No. 193,973, August 7, 1877, Lewis Lillie, "Time-Locks."
54. *See* loose insert to "Memorandum / Corliss #4 Single Movement / Damond [*sic*] #2 Trip[.] C" shipping ledger of Sargent & Greenleaf: "Feb 2nd / 1921 1822."
55. Pat. No. 206,887, August 13, 1878, S. Morris Lillie, "Time-Lock."
56. Hopkins, *Lure of the Lock*, p. 166.
57. *See* Pats. No. 166,632, August 10, 1875, O. F. Pillard, "Time-Lock"; No. 172,629, January 25, 1876, L. Hubbell, "Time-Lock."
58. *See* Pillard advertising circular, undated (ca. 1876), issued by New Britain Bank Lock Co. sales office, 85 Devonshire at Water Street, Boston, and 187 Broadway, New York.
59. *Yale v. City Nat'l Bank of Bridgeport*, Cir. Ct. Mass., decision of Shipman, J., 1877.
60. Pillard advertising circular.
61. Hopkins, *Lure of the Lock*, p. 168.
62. Ibid., p. 178.
63. *Fifty Years of a Successful Industry 1868–1918*, Yale & Towne Manufacturing Co., 1918.
64. Pat. No. 205,275, June 25, 1878, T. F. Keating, "Time-Locks."
65. *Sargent v. Hall Safe & Lock Co.*, Appendix to Exhibits, deposition of James Sargent.
66. *See* Sargent & Greenleaf shipping register, "Memorandum/Regular #4 #4B Timers" (notation inside front cover states movements numbered 1 through 365 were forty-six-hour #4s).

67. Sargent & Greenleaf catalogues list dimensions for the Models 4 and 5.

	1908	1927
Model 4:	4⅝" × 4⅝" × 2¾"	4½" × 4⁷⁄₁₆" × 2¹¹⁄₁₆"
Model 5:	4⁵⁄₁₆" × 4¼" × 2¼"	4⁵⁄₁₆" × 4¼" × 2¼"

68. Pat. No. 206,981, August 13, 1878, E. Stewart, "Time-Lock."
69. Ibid.
70. Howard Ledgers, vol. 4, p. 207.
71. Hopkins, *Lure of the Lock*, pp. 173, 176.
72. Pat. No. 206,146, January 16, 1878, E. Stockwell, "Time-Lock."
73. Statement of Orest Kalba, owner (Anchor Safe Co.), historian (Detroit Safe Co.).
74. Pat. No. 200,312, February 13, 1878, P. F. King, "Time-Lock for Safes."
75. Pat. No. 201,535, March 19, 1878, P. F. King, "Time-Lock."
76. Howard Ledgers, vol. 3, p. 14.
77. Pat. No. 213,809, April 1, 1879, C. E. Chinnock, "Time-Lock."
78. Pat. No. 125,679, April 16, 1872, I. Hertzberg & A. Hertzberg, "Apparatus for Automatically Regulating the Flame of Gas Burners."
79. See, e.g., Pat. No. 3,439 (Impovement in Bee-Hives, S. & J. D. Cope, 1844); Pat. No. 31,322 (Automatic Gas Lighter or Extinguisher, N. S. Manros, 1861); Pat. No. 93,673 (Feed Trough, A. Chambers, 1869)
80. Howard ledgers, vol. 3, p. 41.
81. Ibid., vol. 4, p. 130.
82. Pats. No. 262,095 and 262,103, April 1, 1882, H. F. Newbury, "Time Lock."
83. Howard Ledgers, vol. 5, p. 82.
84. *Yale v. E. Howard Watch Co.*
85. *Ohio State Journal*, June 3, 1879.
86. Dalton Aff. from *Hall v. MacNeale*.
87. Howard Ledgers, vol. 4, p. 153.
88. Ibid.
89. Ibid.
90. Pat. No. 276,383, April 24, 1883, B. F. Flint, "Time Lock."
91. Ibid.
92. Yale sales ledger.
93. Ibid.
94. Howard Ledgers, vol. 6, p. 181.
95. Neff affidavit, USPTO *interference*.
96. Pat. No. 315,612; April 14, 1885; H. Gross; "Lock Mechanism for Safes."
97. Howard Ledgers, vol. 6, p. 244.
98. Hopkins, *Lure of the Lock*, p. 171.
99. Ibid.
100. Howard Ledgers, vol. 8, p. 39.
101. Howard Ledgers, vol. 6, p. 266.
102. Pat No. 321,893; July 7, 1885; H. Gross; "Time Lock."
103. Pats. No. 347,068–347,071, August 10, 1886, F. Sedgewick, "Electromagnetic Permutation Lock."
104. *Bank Locks*, Yale catalogue, 1896, pp. 30–32.
105. Hopkins, *Lure of the Lock*, p. 167.
106. Howard Ledgers, vol. 7, p. 65.
107. Pat. No. 211,409, January 14, 1879, P. F. King, "Time-Lock."
108. Pat. No. 221,789, November 18, 1879, M. A. Dalton, "Time-Lock."
109. Howard Ledgers, vol. 6, p. 99.

CHAPTER 5: The Era of Monumental Security: 1888-1899

1. *Corliss Security System*, Corliss catalogue, undated but post-1894.
2. Ibid.
3. Sargent & Greenleaf shipping ledger, "Memorandum [/] Corliss #4 Single Movement [/] Damond [*sic*] #2 Trip[le] C [/] Feb 2nd [/] 1921 1822."
4. Ibid.
5. Ibid.
6. *Corliss Security System*, Corliss catalogue.
7. Yale sales ledger.
8. "Directions for Using the Yale Triple Movement Time Lock, C," *Yale Bank Lock Catalogue*, February 1889.
9. Ibid.
10. Yale sales ledger.
11. Ibid.
12. Yale sales ledger (notes in red ink that they were exchanged).
13. Ibid.
14. *Lure of the Lock* identifies the Type E in the Mossman Collection as a "Circular Time Lock" that was "made by American Waltham Watch Co.," likely because the earliest Types D and E did not show any attribution to Yale, but the pocket watch movements do clearly note "American Waltham Watch Co."

15. The estimate of 1890 was made by matching dated movement serial numbers to time lock features and then cross-checking the case serial numbers in Yale's sales ledgers.
16. *The Safe Guard*, February 1892, vol. 1, no. 2, p.12 (a monthly publication of the Yale & Towne Mfg. Co., Stamford, Conn.)
17. Yale sales ledger.
18. Emery Stockwell was among the most prolific engineers at Yale and was granted many patents, all of which he assigned to Yale. Consequently, "Stockwell's Patent" is a common inscription among Yale time locks. Later, a Herbert C. Stockwell worked with Emery and then on his own. Their relationship, if any, is unknown.
19. Ibid. This model is stamped "12G," but the Yale sales ledger identifies these time locks as "Triple G" on the page tab and nothing on the page head. There is no known example of this lock with three movements, and "Triple G" may be a bookkeeper's error, mistaking the central dial for a movement. All other aspects of this lock and the records correspond appropriately.
20. "Memorandum [/] Corliss #4 Single Movement [/] Damond [*sic*] #2 Trip[le] C [/] Feb 2nd [/] 1921 1822."
21. Hopkins, *Lure of the Lure*, p. 166.
22. Sargent & Greenleaf catalogue, 1901, p. 10.
23. *See* E. Howard Clock Order Superintendent's Office ledger, vol. 7.
24. Pat. No. 450,293, April 14, 1891, P. F. King, "Time Lock."
25. Ibid.
26. Howard Ledger, vol. 9, p. 185.
27. Yale sales ledger.
28. Shugart et al., *Complete Price Guide to Watches*, 19th edition (Cleveland, Tenn.: Cooksey Shugart Publications, 1999), p. 255.
29. Ibid.
30. Ibid.
31. Pat. No. 508,902, November 14, 1893, M. A. Dalton, "Method of & Apparatus for Controlling & Utilizing Concussion & Applying it to Safe Locks."
32. Although both Yale and Sargent & Greenleaf produced time locks based on an L-size movement, the movements themselves should not be confused or thought to be interchangeable. The "L" designation itself originated from a watch makers' movement size, and the movements used by Yale, made by watch maker E. Howard, did in fact conform to this convention. Sargent & Greenleaf, however, made all its own movements internally and had a background in lock making. Consequently, while Sargent designated one of its modular movements as "L"-sized, this movement was only a rough approximation of the true L.
33. As with the "L" designation, Sargent & Greenleaf's M-size movement only roughly approximated the watchmaking size convention; Yale's M movements, made by E. Howard (serial numbers 1 to 200) and Seth Thomas (all others), were more accurately sized.
34. *See* Yale Lock Mf'g Co., *Yale Bank Locks*, 1908 catalogue, p. 16.
35. Ibid.
36. The name of Hollar's company is not exactly clear, but it was either the "Hollar Company" (see *Scientific American*, April 21, 1906) or the "Hollar Lock Inspection and Guarantee Co." (as shown on an installation plate).
37. Patent Nos. 545,020 and 545,021, August 20, 1895, W. H. Hollar, G. L. Weaver & A. Kennedy, "Electrically Controlled Winding Mechanism for Time Locks." *See also*, Pat. No. 571,319, November 10, 1896, W. H. Hollar & H. W. Pidgeon.
38. *Scientific American*, April 21, 1906, pp. 325, 326.
39. Pat. No. 526,555, September 25, 1894, F. H. Blake, "Time Lock."
40. Howard Ledger, vol. 12, p. 17.
41. Ibid., vol. 11, p. 144; vol. 12, pp. 17, 42.
42. Pat. No. 493,862, March 21, 1893, G. J. H. Goehler, "Time Lock."
43. Pat. No. 520,144, May 22, 1894, A. Kirks, "Time Lock."
44. See *Yale Bank Locks* catalogue, 1908, pp. 16, 17.
45. *See* Yale Lock Mf'g Co., *Yale Bank Locks*, 1908 catalogue, p. 17.
46. Yale sales ledger.
47. *See* Yale Lock Mf'g Co., *Yale Bank Locks*, 1908 catalogue, p. 13.

CHAPTER 6: Locks of the Early Twentieth Century: 1900–1915

1. Pat. No. 769,556, September 6, 1904, E. A. Marsh, "Time Lock."

2. Shugart et al., *Complete Price Guide to Watches*, 19th ed. (Cleveland, Tenn.: Cooksey Shugart Publications, 1999), pp. 235, 237.
3. Yale sales ledger.
4. *See, e.g.,* plaque from Manganese Steel Safe Co. safe #781 (from which the examples shown here came): Nº 781 / Patent / Manganese Steel / Burglar Proof Safe / Hibbard-Rodman-Ely System / Manganese Steel / Safe Cº / New York / Size 5.
5. Yale sales ledger.
6. Ibid.
7. "Memorandum [/] Corliss #4 Single Movement [/] Damond [*sic*] #2 Trip[le] C [/] Feb 2nd [/] 1921 1822."
8. Sargent & Greenleaf, *Sargent & Greenleaf 1929 Catalog.*
9. It is not clear whether there was any relation between Milton Dalton and Harry Dalton, both inventors at the Consolidated Time Lock Co.
10. Pat. No. 775,523, November 22, 1904, H. M. Dalton, "Time Lock"
11. The dials are numbered through 76 but can be wound through to 0, for a total running time of seventy-eight hours.
12. Shugart et al., *Complete Price Guide to Watches*, p. 173.
13. Yale & Towne Mfg. Co., "Directions for using the Yale Duplex Time and Combination Lock," October 1904.
14. Yale sales ledger.
15. Diebold technical paper showing early type of Consolidated plate mounting Yale movements. The whereabouts of this time lock are unknown.
16. Harry C. Miller Collection data sheet for Bankers Dustproof time lock (No. 1040); *see also,* Victor Safe & Lock Co. catalogue (undated but not before 1911), p. 32 (identifying Victor as repair and servicing point for Bankers Dustproof movements).
17. Pat. No. 939,384, November 9, 1909, E. M. Benham, "Time Lock."
18. Ibid., p. 1.
19. Yale sales ledger.
20. Sargent & Greenleaf shipping register for Nos. 4 and 4B time locks (1890–1907), p. 53 (first mention of 4B "hook").
21. Harry C. Miller Collection data sheet for #79 3-mvt. Bankers Dustproof; collection no. 1040.
22. Between the end of Bankers Dustproof time lock production and the beginning of Mosler production, there seems to have been a period during which neither company could provide time locks for Victor Safe Co. During this time, Yale supplied Victor Safe with time locks specially designed to fit in the space of a Bankers Dustproof case, but using a modified Yale L-movement. Fewer than five of these Yale-made time locks from Victor safes are known today.
23. David A. Christianson, CMW, "Mosler Time Lock Movements, Part I," *Horological Times*, September 1989, p. 48.

CHAPTER 7: Alarm Timers
1. Hopkins, *Lure of the Lock*, p. 166.
2. Pat. Nos. 314,282 and 314,283, March 24, 1885, J. B. Young, "Time Lock."
3. Yale sales ledger.
4. Hopkins, *Lure of the Lock*, p. 166.
5. Ibid., p. 167.
6. Ibid.
7. Bankers Electric Protective Company, business correspondence to C. A. Edgerton, cashier, *in re*: then recent use of arc welder to attack safes, November 28, 1898.
8. Howard Ledgers.
9. Shugart et al., *Complete Price Guide to Watches*. E. Howard turned over all operations to Keystone Clock Co. in 1902. There is no record that Keystone produced any time lock or alarm timer movements.
10. Pat. No. 63,158, March 26, 1867, E. Holmes, "Electric Clock."
11. Reissue No. 9,209, May 18, 1880, E. Holmes, "Electric Circuit-Breaking Clock."

GLOSSARY

Automatic *n*. (also **bolt motor**) A mechanism that contains a (usually heavy, wrench-wound) spring designed to automatically throw open the **boltwork** (or, less commonly, the **bolt**) of a safe at a specified time when triggered by a **time lock**.

Bank lock *n*. (also **vault lock**) Generic term for locks designed to secure vaults built into a building or safes whose size and weight render them nearly impossible to practicably move. In safes and vaults of such size, the lock is commonly considered the weakest point of entry and consequently must present such complication and robust construction to survive the most focused attacks.

Bit *n*. 1. The bladelike part of a key that projects from the **shaft** and engages the **tumbler**s when turned, esp. for a lever lock. 2. Each subpiece of a reorderable key that engages a single lever and, when assembled on the **shaft**, collectively form the key's bit.

Bitting the pattern of cuts or indentations in the bit that sets the **tumbler**s.

Bolt *n*. 1. That part in the door or lid of a safe or vault that fastens it to the jamb, preventing the safe from opening when secured. 2. That piece of a lock whose movement comprises the final (un)locking action; the piece whose movement is protected or prevented by the complications of the lock.

Bolt motor *n*. (also **automatic**) A mechanism that contains a (usually heavy, wrench-wound) spring designed to automatically throw open the **boltwork** (or, less commonly, the **bolt**) of a safe at a specified time when triggered by a **time lock**.

Boltwork *n*. The complete mechanism that actuates a number of independent bolts, each of which secures the door, usually thrown by either an exterior handle separate from the lock or a **bolt motor**. Common only in vault doors and larger safes.

Bow *n*. That piece of a key, commonly a loop, that acts as a handle, giving leverage and allowing the **shaft** to be turned in the lock. Bowed keys were standard prior to Linus Yale, Jr.'s "feather key," which did away with both the bow and the shaft, and instead adopted the flat shape common to keys today.

Change key *n*. A key designed to allow the solution to a combination lock to be altered, thus allowing bankers to provide certain people (e.g., tellers) access to the safe or vault who do not have authorization to change the combination.

Combination lock 1. *n*. A lock whose operation requires a specific manipulation of an external control or controls. The security of a combination lock rests in the great number of permutations possible, from thousands in early locks to hundreds of millions in later models.

According to the mathematical definition of "combination" and "permutation," all such locks are in fact permutation locks, i.e., one must not only dial or enter the correct digits of the solution but do so in the correct order. A true combination lock with a solution of 10-20-30 would also open with those numbers in any other order, e.g., 20-30-10. Early in the development of vault locks both terms were used, but eventually (and quite counter to its correct meaning) "combination lock" became the standard term. The last lock denominated a "permutation lock" was patented by Milton Dalton in 1878.

2. *n. arch*. A lever key lock that combined (a) levers whose order could be changed by the user with (b) a key

whose bits could be ordered to correspond to the levers, thus allowing the user to set and reset the order for additional security. The key was commonly carried with the bits incorrectly stacked to avoid inspection of the true key. Like def. 1, such a lock is also technically a "permutation" lock.

Combination tumbler *n.* A circular metal plate with (a) a driving pin that may communicate motion from one tumbler to the next and (b) a strategically placed notch-shaped **gate** that, when properly aligned, will admit the **fence**, so that it no longer **dog**s the **bolt**. A commonly designed combination lock will have one such tumbler for each number in the combination, the dialing of the combination aligning all the notches and thereby opening the lock.

Curb *n.* (also **wheel pack**) A set of combination lock tumblers constructed to allow each tumbler to turn on a common axle while the set is capable of being removed and handled as a single assembly.

Cylinder lock *n.* A cylindrical case containing a keyhole (originally for a flat key but now commonly for a corrugated key) and coil spring–loaded pin **tumbler**s that operates the **bolt** by rotating within a sleeve.

Damascene *v.* To apply a pattern of engine turning to a metal surface, most commonly a set of wavy vertical lines (*n.* damascening).

Dog *n.* Part of a lock mechanism that acts to stop or block the motion of (a) another part of the lock, commonly the bolt, or (b) some moving part outside the lock, as in the case of the action of a **time lock** on a **boltwork**.
 v.t. To act to stop or block the motion of a moving part as described in (a) above.

Dual custody (see also, single custody) *adj.* Regarding any safe with two locking mechanisms (esp., two **combination lock**s), having a **boltwork** requiring that both locks be opened before the safe can be accessed; requiring two keys or combinations, suggesting two people must open the safe together.

Fence *n.* The part of a lock that **dog**s the **bolt** when engaged by the **tumbler**s. The proper key or combination will allow the fence to pass into or through the **gate**, releasing the bolt.

Gate *n.* An opening in a **tumbler**, which the **fence** must move into (in a **combination lock**) or through (in a **lever lock**) in order to release the **bolt**.

Lever *n.* (also lever tumbler) A flat, pivoting tumbler that **dog**s the **bolt**, requiring a key with a **bit** that lifts the lever (to a minimum height in very early models; to a specific height beginning with Barron's 1778 design) before the **bolt** can be thrown. The earliest lever mechanisms had a single lever, but beginning with Barron's design lock makers stacked two or more levers together, significantly increasing security.

Parautoptic *adj.* Obscured from view. Coined by Day & Newell for the 1844 Parautoptic lock design that sought to be unpickable by making the **lever** mechanism impossible to inspect through the keyhole.

Permutation lock *n. arch.,* combination lock.

Shaft *n.* 1. The **spindle** on a **combination lock**. 2. That part of a key that connects the **bit** and the **bow**.

Single custody (see also **dual custody**) *adj.* Regarding any safe with two locking mechanisms (esp., two **combination lock**s), having a **boltwork** requiring that either lock be opened before the safe can be accessed; requiring one of two keys or combinations, suggesting either of two people may open the safe individually, preventing lockouts in the event of a malfunction of one lock or the loss of one key or combination.

Snubber bar *n.* A single sliding piece in a modular-movement time lock that runs horizontally across the case, usually with one vertical projection for each dial pin; it allows any of the movements to release the **bolt**.

Spindle *n.* (also **shaft**) The axle on a **combination lock** that communicates external motion (usually the turning of a numbered dial) through the safe door or the lock case to the interior of the lock; originally a weak point of safe construction. Safecrackers would seek to force the spindle entirely into the safe, pull it entirely out, or drill through it, allowing access to the **combination tumbler**s. Spindles eventually incorporated collared, conical, and offset designs. Offset spindles were also known as "indirect" (referring to the gearwork that transmitted their movements into the lock) and "external" (referring to their location vis-à-vis the case of the lock).

Stump *n.* A small projection that allows one part of a lock to engage another. While the stump(s) in a **lever lock** are sometimes referred to as the **fence**, the **fence** in a lever lock is actually the part of the **tumbler** that the stump strikes when operated by an incorrect key.

Time lock *n.* A lock designed to secure a safe or vault for a predetermined period, rather than until opened with a key or combination. Commonly based on one or more clock or watch movements, time locks gained popularity in the 1870s after a rash of robberies involving kidnapped bank employees.

Tumbler *n.* The part of a lock that defines the required motion to release or throw the **bolt**. See also **combination tumbler, lever**.

Vault lock n. See **bank lock**.

Warded lock *n.* A lock that relies on projections from the case to block the **bit** of an incorrect key for security. Although easily picked and expensive to make, the warded lock was ubiquitous prior to the 1780s, making significant efforts to improve safes inefficient.

Wheel pack *n.* See **curb**.

BIBLIOGRAPHY

Christianson, David A. "Mosler Time Lock Movements." Pt. 1, *Horological Times* (September 1989): 48–51.

———. "Yale & Towne Timelocks." Pts. 1 and 2. *Horological Times* (June 1990): 32–35; (July 1990): 30–33.

———. "72-Hour E. Howard Timelock Movements and Their Variations." *Horological Times* (February 1991): 40–45.

———. "Sargent & Greenleaf Timelock Movements." Pt. 1, *Horological Times* (May 1991): 36–39, 52.

———. "The Yale #1 Timelock and the Stockwell Patent of 1875." Pt. 1, *Horological Times* (September 1991): 46–49.

———. "The Yale #1 Timelock . . . and its E. Howard Movements." Pt. 2, *Horological Times* (October 1991): 20–26.

———. "Sargent & Greenleaf V." *Horological Times* (June 1993): 24–27.

———. "The Victor Banker's Dustproof Timelock Movement." *Horological Times* (April 1991): 48–51.

———. "Hall's Safe & Lock Company's Lock Chronometer." *Horological Times* (October 1993): 28–32.

———, with Walter Bruggeman. "Time Locks: Part I," *Horological Times* (May 1989): 40–43.

———. "A History of the Timelock Industry, Continued. . . to the Present." Pts. 1 and 2. *Horological Times* (April 1994): 34–37; (May 1994): 32–34.

"Chubb's Patent Lock." *Mechanics' Magazine and Register of Inventions and Improvements* (May 31, 1834): 32.

Dalton, Milton A. *History of Fire & Burglar Proof Safes, Bank Locks and Vaults in America and Europe—Useful Information for Bankers, Business Men and Safe Manufacturers.* Cincinnati, Ohio: Franklin Type Foundry, 1874.

This extensive survey of the state of the art of fire- and burglarproof safe making by Dalton may in fact have been one of the most brazen acts of corporate espionage in history. In compiling his *History*, Dalton traveled across the United States and Europe collecting signed statements and notarized affidavits from the then eminent makers in the safe and lock manufacturing industry. We cannot know whether Dalton made clear to the people he interviewed that he was employed by Joseph Hall, a well-known major competitor in the safe and vault lock world, or whether he simply held himself out as writing a two-volume historical work on safes and locks.

Although Dalton's *History* was not organized for easy reference, it is replete with anecdotal evidence and detail about design, manufacture, and the ideas prevalent in the safe making world. In simply reporting the words of competing makers, Dalton gives a picture of an industry still in its youth and exploring a wide array of techniques in search of that ultimate solution to the issues of burglary and fire. We find techniques variously lauded and criticized and in some instances we find makers denying use of certain methods that their competitors attribute to them.

Dalton suggests that a second volume would be forthcoming on the subject of locks, but no trace of any such work has ever been found. In fact, only sixty-six copies of what was designated volume one were ever published and the book was never distributed. It was, however, used by Hall as evidence against competitors in patent litigation. Today, only three copies of Dalton's *History* are known, two in the Mossman Collection and one with the Hall family of Cincinnati, Ohio.

Denison, Edmund Beckett, Adam Edinburgh, and Charles Black. *Clocks and Locks. From the "Encyclopædia Britannica" with a full account of The Great Clock at Westminster.* 1857.

The section on locks (Part II, p. 173) is an excellent reference on the histories and mechanisms of locks until and during the watershed era when this work first appeared. Published following the Great Exposition of 1851, this book reflects major changes that were taking place, changes that are clearly fresh in the writer's mind. *Clocks and Locks* was written at a time when all that was known about locks could still be set forth in a compact volume, and it makes an effort to be exhaustive, discussing everything from the ancient Egyptian locks and warded locks (ethnocentrically dubbed the "Common English Lock") through the then state-of-the-art Magic Key Lock by Linus Yale, Jr. Well illustrated for a survey work of this era, *Clocks and Locks* reproduces many schematics, mostly unattributed but likely from patent drawings, yielding a very usable reference work that combines technically accurate diagrams with a readable text. Unfortunately, the book itself is more of a rarity than many of the locks it describes, and our reference copy from the private library of Dana Blackwell was used with his kind permission.

Dix, Robert W. "The Locks of H. C. Jones." *Journal of Lock Collecting* 33, no. 3 (May 2002): 3–9.

Gibbs, James W. "Horological Treasure Guardians." *Bulletin of the National Association of Watch and Clock Collectors, Inc.* (1965): in three parts, no. 114 (pp. 642–50), no. 115 (pp. 748–66), and no. 118 (pp. 934–49).

Graffeo, Dan. "Consolidated Timelocks." *Safe & Vault Technology* (November 1990): 19–23.

Grieves, Bill, and Lynda Grieves. "Henry Isham's Permutation Lock." *Journal of Lock Collecting* 35, no. 1 (January/February 2004): 9–11.

Haight, A. V. *Yale Bank Locks*. Poughkeepsie, N.Y.: The Yale and Towne Mfg. Company, ca. 1896.

Hennessey, Thomas F. *Locks and Lockmakers of America*. Rev. 3rd ed. Park Ridge, Ill.: Locksmith Publishing Corp., 1976.

In an impressive chronology of the lock industry in Connecticut and throughout the United States, Hennessey offers a survey of everything from patent numbers to builders' hardware. Available through the Lock Museum of America in Terryville, Connecticut, *Locks and Lockmakers of America* devotes an appropriate primary focus to the significant contributions of Terryville and the state of Connecticut. However, Hennessey does not fail to give a reasonable account of the broader montage of people and companies throughout the United States.

Primarily a collection of important chronologies, *Locks and Lockmakers of America* also includes some interesting anecdotal evidence about its subjects, from examples of corporate governance documents to notable labor relations changes. Further, this volume is one of the few sources of information about Alfred North and the New Britain Bank Lock Company, making this an important resource for the lock historian.

Hopkins, Albert A. *The Lure of the Lock*. New York: General Society of Mechanics and Tradesmen, New York, 1928.

Though John Malcolm Mossman's donation of his lock collection and the accompanying endowment to the General Society in 1903 was a welcome gift, the nature, provenance, and import of these pieces was best known by Mossman himself. With Mossman's unexpected death in 1912, the General Society lost not only an invaluable member but also the Rosetta stone to its lock collection.

In the first concerted effort by the General Society to catalogue the contents of the Mossman Collection, Albert Hopkins (a member of the General Society and a respected scientific authority in his own right as associate editor for *Scientific American* magazine) assembled *The Lure of the Lock*, prodigiously subtitled *A short treatise on locks to elucidate the John M. Mossman Collection of locks in the Museum of the General Society of Mechanics and Tradesmen in the City of New York, including some of the "Mossman Papers."* Hopkins's book would come to be a seminal reference for safe and time lock collectors for years to come. Hundreds of photos and diagrams accompany descriptions by specialists from a generation with firsthand experience. As a guide to one of the most complete collections of safe and time locks in the world, *Lure of the Lock* has been the sole contact that a host of avid collectors have had with many of lock collecting's true rarities, and in some cases it represents the only documentation available on some of the subject's unique innovations.

Beyond his description of the items in the Mossman Collection, Hopkins also included interesting excerpts culled from the personal papers of John Mossman, including contemporaneous news reports about the banking industry, the theft "industry," and their coevolution. Sections on the history of lock making survey technology from ancient Egypt through the nascent American safe industry extant during Mossman's childhood, giving a sound history of lock making as well as an excellent basis for understanding the importance of the locks in the collection.

While Mossman's untimely death left many gaps in the understanding of the history of American locksmithing and his collection, the work of dedicated collectors and researchers has steadily filled in details over the decades. *Lure of the Lock* represents the bulk of this detail and will remain a major resource for the serious researcher and collector alike. The current republication of *Lure of the Lock* is available from the General Society of Mechanics and Tradesmen.

Kasper, Mike. "A Burning Question." *Safe & Vault Technology* (February 1999): 24–26.

McOmie, Dave. "MacNeale & Urban Safe." *Safe & Vault Technology* (November 1991): 269–72.

Price, George. *A Treatise on Fire & Thief-Proof Depositories*. London: Simpkin, Marshall, and Co., 1856.

Report of Special Commission of Experts as to means of improving vault facilities of the Treasury Department. U.S. Comm'n on Safe and Vault Construction, Ex. Doc. No. 20, 1893.

INDEX

Page references in *italics* refer to illustrations.

alarm timers, 327
    American Bank Protection Co., *337–39*
    Bankers Electric Protective Assn., *333–36*
    Electric Switch (Yale), 330, *331*
    Diebold, *340–41*
    New Haven Clock Co., 328, *329*
    time lock, differentiated from, 34, 327
American Bank Protective Co., 33
American Bank Protection Co.
    alarm timer, *337–39*
American Institute, 24th Fair of the, 58, 64
American Telephone & Telegraph, 34
American Waltham Watch Co., 172
    Four-movement time lock (Mosler), 324
    Keating's time lock, 192
    Types D and E time lock, 245, 246
    Type G time lock, 248–49
Andrews, Solomon, 23
    changeable bit lever lock, *40–43*
    Davidson's Fire King compared to locks of, 84
    and Schroder bank lock, 146
    snail wheel lock, *46–47*
anti-micrometer device
    Covert, Henry, 96
    Dexter (Herring), *136–37*
    Magnetic (Sargent), 118–19
    Magnetic, rollerbolt (Sargent), 120
    Pull-out dial (Lillie), 127
    Sullivan, T. J., 87
    Yale, Jr., 114
    Yale time and combination lock, *174–75*
Atlee, Samuel, 198
automatic
    bolt motor, 99, 194, 200, 224, *246–47, 248–49, 250–51, 258–59, 262, 264–65, 277, 279, 282–83, 286, 304, 310, 314, 318*
    combination lock, in reference to, 122–23
    time lock, in reference to, 98, 99, 200

Bacon, William W., 96
Bankers Dustproof, 32, 324
    1906 time lock, *314–17*
    origins as Consolidated Time Lock, 32
    relation to Victor Safe & Lock, 32
Bankers Electric Protective Association
    alarm timer, *333–336*
    Bankers Electric Protective Company, name change to, 333–34
Barron, Robert
    1778 lock, *14*, 15
Bates, Mark, 11
    images, courtesy of, *123*
Beard & Bro., 204, 232, 260
    patent litigation, 33
    Pye's time lock identified as Beard's, 64
    Type 1 time lock, *204–05*, 214
    Type 2 time lock, *206–07*
    Type 3 time lock, 206
Betteley, Albert
    safe lock, *70–71*
black powder. *See* explosives
Blackburn, Jacob, 198
Blackwell, Dana J., 11, 182
Blake's Bank Lock Inspection Co.
    Columbian time lock, *284–85*
bolt motor
    Burton-Harris, *250–51, 258–59*
    Dalton, Harry (Consolidated), *304*
    Hall's Automatic, *318–19*
    No. 1 (Yale), *246–47*
    No. 2 (Yale), *248–49*
    No. 3 (Yale), *264–65*
    T-Movement time lock (Yale), 310–11
    See also automatic
Boston Clock Co., 168, 264
Braman, Joseph
    1784 lock, *15*
    200-guinea challenge, 16
    and Schroder bank lock, 146

Brennan, J. B.
    push-key lock, *82*
Briggs, Martin, 96
Briggs–Covert, 96
Burge, John
    Pin Dial, Double (Yale), 162
    Gothic time lock, *144–45*, 148, 151, 170, 200
Burton-Harris & Co.
    bolt motor, *250–51, 258–59*
    Sargent & Greenleaf, acquired by, 258
Butler, E. Rhett, 11
    strongbox, courtesy of, *17*
Butterworth, J. H.
    combination lock, *58–59*
    lock compared to Smith's, 56

Cady, Ira L.
    partner of J. M. Mossman, 9
Cary Safe Company, 286, *287*, 318
change keys. *See* key-changing combination lock
Chicago Edison Co., 330
Chicago Fire of 1871, 29
Chicago Safe & Lock Co.
    Gem time lock, *224–25*, 226
Chicago Time Lock Co.
    Marsh time lock, *298–99*
    Perfection time lock, *226–27*
Chinnok, Charles, 208
Chubb, Jeremiah
    1818 lock, *16*, 46
    detector mechanism, 38, 40, 86
Christianson, David
    Gothic time lock, restoration of, 144
Clemens, Samuel (*aka* Mark Twain), 32
Collins, Lynn, 11
    images, courtesy of, *40–42, 146–47*
combination locks
    150-Number (Hall's Safe & Lock), *140*
    Automatic (Sargent), *122–23*, 150

INDEX

combination locks *(continued)*
    Butterworth's, 56, *58–59*
    change keys. See change keys
    Click lock (Lillie), *83*
    Crescent (Hall's Safe & Lock), *130–31*
    Dalton Triple Guard (Consolidated), 20, *234–37*, 300
    Damon's, *220–21*
    Derby's, *104–05*
    Dexter (Herring), *134–36*, *159*, 172
    Diebold & Kienzle's, *141*
    Double Dial (Yale, Jr.), *114–15*
    Duplex (Yale), *308–09*
    Electro-Magnetic (Yale), 230
    Eureka lock (Dodds), *112–13*
    Evans & Watson's, *67*
    Excelsior (MacNeale & Urban), *116*
    Gear lock (Lord), 101
    Isham-Pillard (New Britain Bank Lock), *124–26*
    K31½ (Yale), *300–01*, 308
    key lock, described as, 40
    LS31 (Yale). See Duplex *above*
    Navigator (La Gard), *343–44*
    Magnetic (Sargent), *118–19, 122*, 150
    Magnetic, rollerbolt (Sargent), *120–21, 122*, 144
    Model 1 (Sargent), *148–49, 174*, 224, 300
    Model 3 time and combination (Sargent), *202–03*
    Perkin's, similarities, *20*
    Premier (Hall's Safe & Lock), 29, *132–33*, 138, 140, 190, 214, 218–19
    Pull-out dial (Lillie), *127*
    Revolving Bolt (Hall's Safe & Lock), *138–39*
    Rodgers's, similarities, 22, *23*
    Sedgewick's Electric (Herring), *230–31*
    Smith's, 56, *57*
    Sullivan's, *87*
    Tilton, McFarland & Co.'s, *117*
    Treasury lock (Dodds), *112–13*
    Weimar's (Herring), *137*
    Yale time and combination lock, *174–75*
Consolidated Time Lock Co., 264, 310, 334
    advertising plaques, 152
    after death of Joseph Hall, 32, 310
    Burton-Harris & Co., early relationship with, 258
    Dalton Concussion Triple time lock, *270–71*
    Dalton Triple Guard time and combination lock, 20, *234–37*
    Dual Guard time lock, 234, *236–37*
    Flint time lock, *218–19*
    founded, 30, 168, 214

Harry Dalton Automatic time lock, *304–05*
Harry Dalton Triple time lock, *306–07*
Infallible time lock, *132–33*
patent litigation, 30, 32
serial numbering, 167–68
Two-Movement time lock, *214–17*
Victor Safe & Lock Company, acquisition by, 32, 314
Woolley's Patent time lock, *228–29*
Corliss Safe Co., 240
    Planet safe, 240, *242*
    Spherical safe, 240, *243*
Corliss, William, 240, 243
Covert, Henry, 96
    combination lock, *96–97*
Cross, Dean, 11
    images, courtesy of, *209*

Dalton, Harry, 304–05, 306–07
Dalton, Milton, 30
    Concussion Triple time lock, *270–71*
    History of Fire and Burglar Proof Safes, 30
    patent of permutation lock, 20
    Triple Guard time and combination lock, 20
    Two-Movement time lock, *212–13*
Damon, George, 220
Damon's Bank Lock Co.
    bank lock, *220–21*
Damon Safe Co., 254
Davidson, J.
    Fire King Bank lock, *84–85*
    MacBride &, 84, 87
Day, Newell & Miner, 70
Day & Newell
    Parautoptic lock, *53–55*, 146
    Parautoptic lock, Davidson's Fire King compared, 84
    similarity of Jones's mechanism, 68
Day's key lock, 84, *86*
Delano, Jesse, 18
Derby, Lyman
    combination lock, *104–05*
Detroit Safe Co., 202–03
Deuber Watch Co., 298
dial lock, 58
    See also combination locks
Diebold & Kienzle
    combination lock, *141*
Diebold Safe & Lock Co., *272*, 306, 344
    alarm timer, *340–41*
    automatic bolt motor, *280–83*
    Chicago Time Lock and, 298–99
    Diebold & Kienzle, as successor to, 141
    Harry Miller, 10

Four-movement time lock, *291–93*
Tisco prototype time lock, *256–57*
Tisco time lock, *288–90*
Type 1 time lock, *280–83*
disconcerter, 106
Dodds-MacNeale-Urban, 112
dynamite. See explosives

E. Howard Clock Company
    American Bank Protection Co., 337, 339
    Bankers Electric Protective Assn., 332–33, 335
    Beard & Bro., 206
    Blackwell, Dana, and (as Howard Watch and Clock Products Co.), 11
    Blake's Bank Lock Inspection, and, 284–85
    Chicago Safe & Lock, and, 224
    Chicago Time Lock, and, 226
    Consolidated Time Lock, and, 228, 236
    Diebold, and, 292
    Hall's Safe & Lock, and, 166–67
    Holmes Time Lock, and, 33, 208–09, 210
    Keystone Watch Case Co., takeover by, 265, 280, 316, 337, 340
    L-movement, 267
    Mosler, and, 232, 260
    patent litigation, 33, 168
    Pye's time lock, and (as Howard & Davis), 64
    Stewart, and, 198
    time lock movements, 26, 172, 191, 212, 280
    Yale, and, 164, 200, 209, 222, 264, 276, 284, 330
Eastman, Joseph, 168
Edison, Thomas, 34
electricity
    effect of electrification, 32, 327, 343
    Holmes Electric time lock, *208–09*
    Sedgewick's Electric combination lock, *230–31*
Elgin Watch Co., 304, 306
Elward Safe Key Lock, 78
Elwell, 78
Ely Norris Safe Co., 304–05, 310, 320
England
    mercantilist policy and U.S. lock industry, 14
Evans & Watson
    letter combination lock, *67*
    Lillie's Click lock, letter lock compared to, 83
Evans, David, 67

INDEX

exhibitions
    Exposition universelle de Paris, 284
    Great Exhibition of 1851, 46, 60
    Paris Exhibition of 1867, 114
    Twenty-fourth Fair of the American Institute, 58, 64
    World's Columbian Exposition, 284
explosives
    Grasshopper lock (W. Hall), and, 60
    Model 2 (Sargent), and, 158
    research by Diebold, 288–89
    research by Newbury, 180, 182, 210–11
    Nock's lock, and, 21
    Peanut Key lock (Yale, Sr.), and, 88
    safe cracking, introduction into, 22
    safe design, and, 108, 312
    Triple B time locks (Sargent), and, 252
Exposition universelle de Paris, 284

factory organization, *111*
Federal Reserve System, 343
fire
    Chicago Fire of 1871, 29
    New York fire of 1835, 50
    Park Row fire, *31*
    at Patent and Trademark Office, 18, 21, 22, 44
fireproof safe
    compared to burglarproof safe, 50
    Fire King (Davidson), *84–85*
    Evans & Watson, makers, 67
    Sterns & Marvin's lock, *80*
    Rich & Co., makers, 50
    Tilton, McFarland & Co.'s lock, *117*
Flint, Benjamin
    time lock, *218–19*
Fraser, Edwin and Elizabeth, 11

G Attachment. *See* Gesswein Attachment
Gayler, C. J., 38
    Gayler's Bank Lock, 38, *39*
Gem Time Lock Co., 224
General Society of Mechanics and Tradesmen of the City of New York
    membership of J. M. Mossman, 10
Gesswein Attachment, *264–65*, 291, 303
Goehler, George, 286
grasshopper lock, *60–61, 62–63*
    Brennan's lock, compared to, 82
    Peanut Key lock (Yale, Sr.), compared to, 88
Great Depression, 161, 343
Great Exhibition of 1851, 46
Greenleaf, Colonel Halbert
    and Linus Yale, Jr., 24
Gross, Henry
    150-Number combination lock, 140

Chicago Safe & Lock Co., 224
Chicago Time Lock Co., 226
Crescent combination lock, 130
Gem time lock, 224, 226
Hall Premier combination lock, 29
Hall's Safe & Lock Co., 166, 224
Hall's Safe & Lock Co.'s time lock, 166
Perfection time lock, 226
guild system, effect on locksmithing, 14
gun powder. *See* explosives

Hall family, *33*, 305
    images, courtesy of, *29, 33*
Hall, Joseph L., IV, 11
Hall, Joseph Lloyd, *29*, 176, 305, 310
    Consolidated Time Lock Co., formation of, 30, 168, 214
    death of, 32
    motto, 30
    on Holbrook's Automatic, 98
    patent litigation, 29, 32
Hall, T. Cartwright, 11
Hall, William
    Grasshopper lock, *60–61*
    Changeable Bit key lock, *78–79*
Hall's Safe Co., 32, *318–19*
    bolt motor, *318–19*
    trade name litigation vs. Herring-Hall-Marvin, 32
Hall's Safe & Lock Co., 29, 178, 305
    150-Number combination lock, *140*
    after death of Joseph Hall, 32
    Automatic (Holbrook), offered for sale, 98
    Crescent combination lock, *130–31*
    Dalton, Milton, and, 30
    Dalton's Two-Movement time lock, *212–13*
    Great Chicago Fire, 29
    Infallible time lock, *132–33*, 172, *190–91*, 228
    MacNeale & Urban, versus, 30
    predecessor companies, 29
    Premier safe lock, 29, *132–33*, 138, 140, 190, 214, *218–19*
    Revolving Bolt combination lock, *138–39*
    serial numbering of time locks, 167–68
    time lock, *166–69*
    Victor Safe & Lock, acquisition by, 32
    Woolley's Fluid time lock, *176–77*
Hampden Watch Co., 298
Harig & Stoy
    lever lock, *81*
Harvard Clock Co., 168
Hayward, Samuel W., 98, 100
Hendrickson, E. M., 62
Hendrickson Safe Co.
    bank lock, *62–63*

Hennessey, Thomas F., Sr., 11
Herring, Silas C., 8, 60, 68
Herring & Co., 60, 68
    Champion safe, *159*
    Dexter combination lock, *134–36, 159*, 172
    Infallible time lock, *172–73*, 192
    patent litigation, 172, 192
    Sedgewick's Electric combination lock, *230–31*
Herring Farrel & Co., 137
Herring, Farrel & Sherman, 136
Herring-Hall-Marvin Co., 32, 104
    Premier safe lock, 132
    trade name litigation vs. Hall's Safe, 32
    See also Marvin Safe Co.
Herring Lock Co., 137
Hertzberg, Abraham, 208
Hertzberg, Isaac, 208
Hibbard-Rodman-Ely Safe Co., 300, 308
History of Fire & Burglar Proof Safes, 30
Hobbs, Alfred C.
    Bramah lock, picked, 16, 76
    Chubb lock, picked, 16, 76
    Jenning's Snail Wheel lock, picked, 46
    production of improved Parautoptic in England, 54
hobnail safe, 18, *19*, 22
Holbrook, Amos
    Automatic lock, 26, 64, *98–100*, 145, 148, 244
Hollar Company, 34, 35
    time lock, *276–79*
Holmes, Edwin
    and Bell Telephone, 34
    introduction of electric alarms, 32, *34*
    patent litigation, 27, 33, 168
    reissued patents, 33, 338
    Holmes Electric time lock, 208–09, 245
    Holmes Electric pendulum, 209
    Holmes New Electric time lock, 209–11
Holmes Burglar Alarm & Telegraph, 27
    founded (as Holmes Burglar Alarm Co.), 32
Holmes Time Lock Co., 255
    Holmes Electric, *208–11*, 245, 260, 295, 298
Howard & Davis, 64. *See also* E. Howard Clock Co.
Howard Watch and Clock Products Co., 11. *See also* E. Howard Clock Co.
Hubbell, Laporte
    Gothic time lock, *144*, 145
    Pillard time lock, 186
    Stockwell Demonstration time lock, 170

INDEX

Illinois Watch Company, 314, 324
Isham, Henry
    Isham-Pillard bank lock, *124–26*
    Key Register lock, *124–26*
    Lillie's Click lock, Key Register lock compared to, 83
    Permutation lock, 90, *91*

J. & J. Taylor, 198
J. H. Schroder & Co. *See* Schroder, J. H.
J. M. Mossman Company, 9
Jennings, 46
Jones, H. C.
    lever lock, *68–69*

Kalba, Orest, 11
Keating, Thomas F., *144*
    and Gothic time lock, 144
    Holbrook's Automatic, purchase of rights, 98
    partner of J. M. Mossman, 9
    time lock, *193*
key lock. *See* lever lock, pin tumbler lock
key-changing combination lock
    150-Number (Hall's Safe & Lock), *140*
    Damon's, *220–21*
    Dexter (Herring), *134*
    Excelsior (MacNeale & Urban), *116*
    Magnetic (Sargent), *118–19*
    Magnetic Rollerbolt (Sargent), *120*
    Pillard, 128
    Sullivan, T. J., *87*
    Yale time and combination lock, *174–75*
Keystone Watch Case Co., 265, 280, 316, 337, 340
King, Phinneas, 64, 204, 232, 260–61
Kingsmill, Joseph, 11
Kirks, 292
Kodas, Andrew and Daniel, 11
Krenkel, K.
    key lock, *92–93*

La Gard, Inc.
    Navigator lock, 343–44
L'Art du serrurier, 15
lever lock
    Andrews's, *40–43*, *46–47*, 146
    Barron's, *14*, 15
    Betteley's, *70–71*
    Day's, 84, *86*
    Double Treasury (Yale, Jr.), *106–07*
    Fire King (Davidson), *84–85*
    described as combination lock, 40
    described as permutation lock, 90
    Gayler's, 38, *39*
    Hall's Changeable Bit, *78–79*
    Harig & Stoy's, *81*

Isham's Permutation, 90, *91*
Jones's, *68–69*
Krenkel's, *92–93*
Magic Key (Yale, Jr.), *72–74*
Miller's, *102–03*
Newell's, *44–45*
Nock's, 20, *21–22*, 146
Parautoptic (Day & Newell), *53–55*, 144
Patrick's, *75*
on Premier bank lock, 132
Schroder's, *146–47*
Tilton, McFarland & Co.'s, *117*
Wilder's, 50, *51*
Lillie, Lewis
    Click lock, *83*
    fireproof safe lock, 52
    Model 1 time lock, *178–79*
    Model 2 time lock, *184–85*
    patent litigation, 30, 178, 184
    Pull-out dial lock, *127*
    and Schroder bank lock, 146
Lillie, S. M., 184
Little and Burge
    Pin Dial, Double (Yale), 162
Lockmasters Security Institute, 10, 23
Lord, John P.
    Gear lock, 101
Loriot & Orstrom, 184
Lure of the Lock
    on Covert's combination lock, 96
    on Davidson's Fire King lock, 84
    on Day's bank lock, 84
    on Diebold's Tisco Safe prototype time lock, 256
    on Krenkel's key lock, 92
    on Mosler's 1887 time lock, 232
    on New Haven Clock Co. alarm timer, 328
    on Newbury's Model, 192
    on Rodgers's 1830 lock, 21
    on W. Hall's Changeable Bit key lock, 78
    on Sargent & Greenleaf's Model 2[5] time lock, 155
    on Sargent & Greenleaf's time lock priority, 96
    on Yale & Towne's Electric Switch, 330

MacBride, 84
MacBride & Davidson, 84, 87
MacNeale & Urban
    Dual Guard time lock, use of, 235–36
    Excelsior combination lock, *116*
    patent litigation with Hall's Safe & Lock, 30
    Type BB time lock, use of, 245
Manganese Steel Safe Co., 264, 300, 308
Marietta Torpedo Company, 289
Marsh, Ernest, 298

Marvin & Co., 104. *See also* Marvin Safe Co.
Marvin Safe Co.
    factory organization, *111*
    fireproof safe lock, 52
    and Park Row fire, *31*
    as predecessor Marvin & Co., 104
    as predecessor Sterns & Marvin, 80
    vault door sample, *108–10*
masked robbery, 25, 143, 190
micrometer, 25, 58, *118*, 127, 129, 131, 174, 230. *See also* anti-micrometer device
Milford Journal, 98, 100
Miller, Harry, 10
    images, courtesy of Harry Miller Collection, *209*, *270*
Miller, J. Clayton, 10
Miller, L. H.
    bank lock, *102–03*
Morgan, John Pierpont, 239
Mosler Bahmann, 324
Mosler Safe Co., 324
    Four-movement time lock, *324–25*
Mosler Safe & Lock Co., 178, 206
    1887 time lock, *232–33*
    Calendar time lock, 232, *260–61*
    Mosler Safe Co., relation to, 324
Mossman, John Malcolm, 8–10, *9*, 274
    member of General Society of Mechanics and Tradesmen, 10
    partner of Thomas Keating, 9, 193
    lock collection, 10, 14
Mossman, Malcolm, 8
Mossman, William, 9
Munger, Lyman, 150

National Safe Co., 322

Neff, E. W., 226
New American Permutating Lock, *53–55*
New Britain Bank Lock Co.
    Isham's permutation lock, 90
    Isham-Pillard lock, *124–26*
    patent litigation, 30
    Pillard combination lock, *128–29*
    Pillard time lock, *186–89*, 245
New Haven Clock Co., 328
Newbury, Henry F.
    explosives research, 180, 182, 210–11, 252, 255
    Model time lock, *192*
Newell, Robert
    1838 patent, *44–45*
    Parautoptic Lock, *53–55*
    Parautoptic lock, Davidson's Fire King compared, 84
    at Twenty-fourth Fair of the American Institute, 64

365

nitroglycerin. *See* explosives
Nobel, Alfred, 22
Nock, Joseph
    lock from Second Bank of the United States, 20, *21–22*
    and Schroder bank lock, 146
North, Frederic
    and New Britain Bank Lock Co., 124, 186

Parker Brothers, 187
patent litigation, 26
    Beard & Bro., 33
    court system, overview of, 30
    Damon's Bank Lock Co., 30
    E. Howard Clock Co., 33
    Gross, Henry, 224
    Hall v. MacNeale & Urban, 166
    Hall's Safe & Lock Co., 30, 166–67
    Herring & Co., 172, 192
    Holmes Burglar Alarm & Telegraph, 27, 33, 192
    Lillie, Lewis, 30, 178, 184
    MacNeale & Urban, 30, 166
    Newbury, Henry, 252
    Pillard, Oliver, 128
    Pye's time lock mentioned, 64
    Sargent & Greenleaf, 26, 27, 30, 128, 144, 145, 172, 184, 192, 224
    Stockwell Demonstration time lock (Yale), 170
    Supreme Court of the United States, 31, 32
    Yale Lock Mfg. Co., 26, 27, 30, 33, 144, 145, 178, 184, 186, 192
    Yale & Towne Mfg. Co., 224
    Yale Lock Mfg. Co. v. Berkshire Nat'l Savings Bank, 27, 30, 32, 166, 172
patent model
    Gothic (Burge), *144–45*
    Model 1 (Sargent), *148–49*
    Newel's, *44–45*
    Yale time and combination lock, *174–75*
Patrick, Robert N.
    push-key lock, *75*
Peerless Lock Co., 141
Perkins, J.
    1813 lock, 18, *20*
    lock compared to D. M. Smith's, 56
permutation lock, 20, 90, *91*. *See also* combination lock
Perry, Stewart
    and New Britain Bank Lock Co., 124, 126
Pillard, Oliver
    combination lock, *128–29*
    and Crescent combination lock, 131
    Isham-Pillard lock, *124–26*
    time lock, *186–89*, 245

pin tumbler lock
    Back-Action (Yale, Sr.), *76–77*
    Bramah's, *15*, 144
    feather key (Yale, Jr.), *114–15*
    Peanut Key (Yale, Sr.), *88–89*
    Perkin's, similarities, 20
    Quadruplex (Yale, Jr.), 48, *49*
    Stansbury's, *18*
push-key lock
    Brennan's, *82*
    Grasshopper (W. Hall), *60–61*
    Hendrickson's, *62–63*
    Patrick's, *75*
    Sterns & Marvin's, *80*
Pye, Francis
    time lock, *64–66*, 148

Rich & Co., 50
Ritchey, H., 68
Rodgers, G. A.
    1830 lock, 21, *23*
    lock compared to D. M. Smith's, 56
Rutherford
    1831 time lock patent, 26, 66, 143, 144

salamander, as fireproof safe symbol, 50, 52, 75
sample, sales and exhibition
    150-Number (Hall's Safe & Lock), *140*
    Crescent (Hall's Safe & Lock), *130–31*
    Gothic (Burge), *144–45*
    Infallible (Hall's Safe & Lock), *190–91*
    Model 1 (Sargent), 148
    Revolving Bolt (Hall's Safe & Lock), *138–39*
    Type G (Yale), *248–49*
    vault door (Herring), *108–10*
    Yale time and combination lock, *174–75*
Sargent, James, 24, *25*, 26, 178
    Automatic, 25, *122–23*, 150, 174
    change key feature, 118, 128
    and Covert, as coinventor, 96
    Magnetic, 25, *118–19*, 122, 150, 174
    Magnetic, rollerbolt, *120–21*, 122, 144
    micrometer of, 25, 58, *118*, 127, 129, 131, 174, 230
    Model 1 time and combination lock, 148, 174
Sargent & Greenleaf, 166, 291, 343
    advertising plaques, 152
    Automatic combination lock, 56, 87, *122*
    Burton-Harris & Co., acquired, 258
    Cary Safe Automatic, *318–19*
    Cary Safe time lock, 286, *287*
    Cleoh time lock, *322–23*
    Contract Respecting Time Locks with Yale Lock Mfg. Co., 27, 28, 274, 291, 330

Corliss Model 4 time lock, *240–41*
    founded, 25
    incorporation (to Sargent & Greenleaf Co.), 160, 254
    Magnetic, *122*
    Model 1, *148–49*, 174, 202–03, 224, 300, 322
    Model 2, *26*, *150–61*, 162, 180
    Model 3, *180–83*
    Model 3, bottom-release, *258–59*
    Model 3 w/combination lock, *202–03*
    Model 3A, *258–59*
    Model 4, *195–97*, 302
    Model 4 Corliss, *240–41*
    Model 4B, *322–23*
    Model 5, 197
    Model 6, *302–03*
    Model O, *273*
    patent litigation, 26, 27, 30, 128, 144, 145, 172, 184, 192, 224
    reorganization (to Sargent & Greenleaf, Inc.), 161
    as Sargent & Greenleaf, Inc., 10, 148, 155
    time lock priority, 98
    Triple A, *250–51*
    Triple B, 250, *252–55*
    Triple C, 250
    Triple D, 250
    Yale time and combination lock, *174–75*
    Variant Model 2, *194*
Savage's time lock patent, 66
Schroder, J. H.
    bank lock, *146–47*
Scientific American
    on Hollar time lock, 278
    on safe construction, 35
Sedgewick
    Electric combination lock, *230–31*
Seth Thomas Clock Co., 280, 330, 332
    American Bank Protection Co., and, 337
    Diebold, and, 340
    Yale, and, 264, 265, 267, 276, 300, 308, 312, 320
    See also Thomas, Seth
Shipman, Judge Nathaniel
    effect of early decisions, 27, 186
    overturned by Supreme Court, 32
Shoop, James 11
    photos, courtesy of, *280–81*, *288–89*
skeleton key, 14
Smith, D. M.
    1846 lock, 56, *57*
Sommers, Steven, 11
South Bend Watch Co., 304, 318
Stansbury, A. O.
    1807 lock, *18*
    similarity of key to Jones's, 68

INDEX

Sterns & Marvin
  fireproof safe lock, 80
  See also Marvin Safe Co.
Stewart, Edward, 178
  time lock, 198–99
Stewart Time Lock Co., 198
Stockwell, Emory, 164, 248
  Demonstration time lock (Yale), 170
  Pin Dial, Double (Yale), 162, 200
strongbox, 16–18, 17
Sullivan, T. J.
  combination lock, 87
Supreme Court of the United States, 31, 32

Terwilliger, William
  employer of J. M. Mossman, 8
  fireproof safe lock, 52
Thomas, Seth
  Model 1 (Lillie), 178
  Model 2 (Sargent), 158
  Model 2 (Lillie), 184
  Pin Dial, Double (Yale), 162, 164
  Stockwell Demonstration (Yale), 170
  See also Seth Thomas Clock Co.
Tilton, McFarland & Co.
  fireproof safe lock, 117
time locks, 26, 143
  1887 (Mosler), 232–33
  1906 (Bankers Dustproof), 314–17
  alarm timer differentiated, 34
  Automatic (Holbrook), 26, 64, 98–100, 144, 148, 244
  Automatic (Yale), 200–01
  Calendar (Mosler), 232, 260–61
  Cary Safe (Sargent), 286, 287
  Cary Safe Automatic (Sargent), 318–19
  Cloeh (Sargent), 322–23
  Columbian (Blake's), 284–85
  Corliss Model 4 (Sargent), 240–41
  Dalton Concussion Triple (Consolidated), 270–71
  Dalton Triple Guard (Consolidated), 20, 234–37, 300
  Diebold Four-movement, 291–93
  Dual Guard (Consolidated), 234, 236–37
  Duplex (Yale), 308–09
  explosives, susceptibility to, 210
  Flint (Consolidated), 218–19
  Four-movement (Mosler), 324–25
  Gem (Chicago Safe & Lock), 224–25, 226
  Gothic (Burge), 144–45, 148, 151
  Hall's Safe & Lock Co.'s, 166–69
  Harry Dalton Automatic (Consolidated), 304–05
  Harry Dalton Triple (Consolidated), 306–07
  Hollar's, 34

Holmes Electric, 33, 208–09, 295, 298
Holmes New Electric, 209–11
Infallible (Hall's Safe & Lock/Consolidated), 132–33, 172, 190–91
Infallible (Herring), 172–73, 192
K31½ (Yale), 300–01, 308
Keating's, 192
L-Movement time lock (Yale), 312–13
LS31 (Yale). See Duplex
Marsh (Chicago Time Lock), 298–99
Model 1 (Sargent), 148–49, 174, 202–03, 224, 300, 322
Model 1 (Lillie), 178–79
Model 2 (Sargent), 26, 150–61, 162, 180
Model 2 (Lillie), 184–85
Model 3 (Sargent), 180–83
Model 3 bottom-release (Sargent), 258–59
Model 3 time and combination (Sargent), 202–03
Model 3A, 258–59
Model 4 (Sargent), 195–97, 302
Model 4, Corliss (Sargent), 240–41
Model 4B (Sargent), 322–23
Model 5 (Sargent), 197
Model 6 (Sargent), 302–03
Model O (Sargent), 273
Newbury's Model, 192
Perfection (Chicago Time Lock), 226–27
Pillard (New Britain Bank Lock), 186–89, 245
Pin Dial, Double (Yale), 158, 160, 162–65, 167, 187, 200, 209, 236
Pin Dial, Single, 222–23
Pye's, 64–66, 148
Quad M (Yale), 294–95
Quad N (Yale), 276, 278, 284, 294
Rutherford, earliest patent by, 26, 143, 144
Savage's patent, 66
Stewart's, 198–99
Stockwell Demonstration (Yale), 170–71
T-Movement automatic (Yale), 310–11
Tisco (Diebold), 288–90
Tisco prototype (Diebold), 256–57
Triple A (Sargent), 250–51
Triple B (Sargent), 250, 252–55
Triple C (Sargent), 250
Triple D (Sargent), 250
Triple K (Yale), 268–69, 280
Triple L (Yale), 262–27, 268, 269, 280, 295, 300
trust agreement between Sargent and Yale regarding, 27
Two-Movement (Dalton), 212–13
Two-Movement (Consolidated), 214–17
Type 1 (Beard & Bro.), 204–05, 214
Type 1 (Diebold), 280–83

Type 2 (Beard & Bro.), 206–07
Type 3 (Beard & Bro.), 206
Type B (Yale), 186, 244–45
Type BB (Yale), 245
Type C (Yale), 244–45
Type D (Yale), 186, 244–47, 248, 262
Type DD (Yale), 246
Type E (Yale), 186, 244–47, 248, 262
Type EE (Yale), 246
Type G (Yale), 248–49
Variant Model 2 (Sargent), 194
Woolley's Fluid (Hall's Safe & Lock), 176–77
Woolley's Patent (Consolidated), 228–29
Y-261 (Yale), 320
Y-361 (Yale), 320, 321
Yale time and combination, 174–75
Towne, Henry R.
  J. M. Mossman, eulogized by, 9
  and Linus Yale, Jr., 24
Twain, Mark. See Clemens, Samuel

United States, Second Bank of, 20
United States Patent and Trademark Office
  1836 fire, 18, 21, 22, 44
  fireproofing patents, 50
  Gothic time lock, 144
  litigation, 27
United States Treasury Department
  use of Dodds Eureka lock, 112
  use of Holmes Electric, 33, 208, 210
  use of Yale Double Treasury, 106

Victor Safe & Lock Company, 312, 314, 324
  Bankers Dustproof Time Lock Co., and, 314
  Consolidated Time Lock, acquisition of, 32, 314

W. B. Dodds Co.
  Lord's Gear lock, similar design, 101
  and MacNeale & Urban, 116
  Treasury lock, 112–13
war, effect on industry, 343, 344
warded lock, 14
Watson, James, 67
Watson & Son, 67
Western Union, 34
Weimar, Jacob, 137
  combination lock, 137
  Infallible time lock, 172
Wilder, George, 50
  Salamander Safe, 50, 51
Woolley, Edward J.
  Fluid time lock, 176–77
  Patent time lock, 228–29

367

INDEX

World's Columbian Exposition, 284
World's Safe Co., 83

Yale, Charles O.
    Infallible time lock (Herring), 172
Yale, Linus, Jr., *24*
    $3000 challenge, *54*
    death of, 222
    Double Dial bank lock, *114–15*
    Double Treasury lock, *106–07*
    feather key lock, *114–15*
    Keating time lock, 193
    key designs similar to Betteley's, 70
    Krenkel's lock compared to Magic Key lock, 92
    Magic Key lock, 72, 74
    Parautoptic lock, ability to pick, *54*
    and Stockwell, Emory, 162
Yale, Linus, Sr., 24
    Back-Action lock, 76
    key designs similar to Betteley's, 70
    Peanut Key lock, *88–89*
    Quadruplex lock, 48

Yale Lock Mfg. Co., 24, 166, 178
    advertising plaques, 152
    Automatic time lock, *200–01*
    Contract Respecting Time Locks with Sargent & Greenleaf, 27, *28*, 274, 291, 330
    Gothic time lock, purchase of, 144
    Holbrook's Automatic, purchase of rights to, 98
    Keating's time lock, *192*
    patent litigation, 26, 27, 30, 33, 168, 178, 186, 192
    Pin Dial time lock, Double, 158, 160, *162–65*, 167, 187, 200, 236
    Stockwell Demonstration time lock, *170–71*
Yale & Towne Mfg. Co., 222, 274, 280
    Duplex time and combination lock, *308–09*
    Electric Switch alarm timer, 330, *331*
    Electro-Magnetic combination lock, 230
    K31½, *300–01*, 308
    L-Movement time lock, *312–13*
    LS31. *See* Duplex
    No. 1 automatic, *246–47, 262, 277, 279*
    No. 2 automatic, *248–49*
    No. 3 automatic, *264–65*
    patent litigation, 224
    Pin Dial, Single, *222–23*
    Quad M, *294–95*
    Quad N, 276, 278, 284, 294
    Sextuple time lock, 246
    T-Movement automatic, *310–11*
    Triple K, *268–69*, 280
    Triple L, *262–67*, 268, 269, 280, 295, 300
    Type B, 186, *244–45*
    Type BB, 245
    Type C, *244–45*
    Type D, 186, *244–47*, 248, 262
    Type DD, 246
    Type E, 186, *244–47*, 248, 262
    Type EE, 246
    Type G, *248–49*
    Y-261, 320
    Y-361, 320, *321*
York Safe & Lock Company, 9, 254, 320
Young, Jackson B., 328

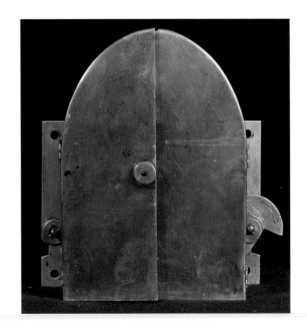